TEACHER EDITION

TEACHING MATH USING LEGO® BRICKS

BASIC MEASUREMENT

USING LEGO® BRICKS

Dr. Shirley Disseler

COMPASS

Basic Measurement Using LEGO® Bricks—Teacher Edition

Published by Brigantine Media/Compass Publishing
211 North Avenue, St. Johnsbury, Vermont 05819

Cover and book design by Anne LoCascio
Illustrations by Curt Spannraft

Brigantine Media/Compass Publishing
211 North Avenue
St. Johnsbury, Vermont 05819
Phone: 802-751-8802
Fax: 802-751-8804
E-mail: neil@brigantinemedia.com
Website: www.compasspublishing.org
www.brickmath.com

ORDERING INFORMATION
Quantity sales
Special discounts for schools are available for quantity purchases of physical books and digital downloads.
For information, contact Brigantine Media at the address shown above or visit
www.brickmath.com

Individual sales
Brigantine Media/Compass Publishing publications are available through most booksellers.
They can also be ordered directly from the publisher.
Phone: 802-751-8802 | Fax: 802-751-8804
www.compasspublishing.org
www.brickmath.com
ISBN 978-1-9384067-8-2

CONTENTS

DEDICATION

Dedicated to all those who teach! Teaching requires dedication, patience, and a mind for engaging materials.

INTRODUCTION

Measurement can be a difficult concept for young learners in grades K-5 to master as they develop their understanding of both metric and customary measures. But measurement is one of the most useful mathematic strands in everyday life. The National Council of Teachers of Mathematics defines measurement as "the assignment of a numerical value to an attribute or characteristic of an object" (NCTM, 2000, p.1).

Students begin to understand the basic concept of measurement by first investigating and comparing the physical attributes of objects. Students should be provided with adequate opportunities to practice measurement and develop the skills to solve any measurement problem. Measurement can be combined with more procedural math concepts such as addition, subtraction, multiplication, division, and fractions to give students an application base for these number skills (Burns, 2007).

The National Council of Teachers of Mathematics (2000) sets forth the following expectations for measurement that students should master within each grade band:

Grades K-2

- recognize the attributes of length, volume, weight, area, and time
- compare and order objects according to these attributes
- understand how to measure using nonstandard and standard units
- select an appropriate unit and tool for the attribute being measured

Boaler, Jo. *Mathematical Mindsets: Unleashing Students' Potential through Creative Math, Inspiring Messages and Innovative Teaching.* San Francisco: Jossey-Bass, 2016.

Burns, Marilyn. *About Teaching Mathematics: A K-8 Resource.* 3rd ed. Sausalito, CA: Math Solutions Publications, 2007.

National Council of Teachers of Mathematics. *Principles and Standards for School Mathematics.* Reston, VA: NCTM, 2000.

Grades 3-5

- understand such attributes as length, area, weight, volume, and size of angle and select the appropriate type of unit for measuring each attribute
- understand the need for measuring with standard units and become familiar with standard units in the customary and metric systems
- carry out simple unit conversions, such as from centimeters to meters, within a system of measurement
- understand that measurements are approximations and how differences in units affect precision
- explore what happens to measurements of a two-dimensional shape such as its perimeter and area when the shape is changed in some way

Grades 6-8

- understand both metric and customary systems of measurement
- understand relationships among units and convert from one unit to another within the same system
- understand, select, and use units of appropriate size and type to measure angles, perimeter, area, surface area, and volume

Van de Walle, John A., Karen S. Karp, and Jennifer M. Bay-Williams. *Elementary and Middle School Mathematics: Teaching Developmentally.* 8th ed. Boston: Pearson, 2012.

The term "attribute," meaning a quality or characteristic to be measured (e.g. length, distance, volume, time, etc.), is an unfamiliar one to young learners. To develop understanding, students should be encouraged to measure many objects using both non-standard units and standard units. One cannot develop a full understanding of measurement by simply using a ruler. Many different attributes can be measured, and each one must be explored in turn (liquid volume, mass, time, money, etc.). Students should not merely be given tables and charts to memorize, because this will emphasize the procedural aspects of conversion rather than building their understanding of the underlying concepts. According to John Van de Walle et al. (2012), "The determination of measures of length, height, size, or weight is an important use of number for the young child. Measurement involves meaningful counting and comparing (number relationships) and connects number to the world in which the child lives."

Representational tools can help young students to develop knowledge of underlying concepts and prepare them to engage with the procedural aspects of conversion. For example, students learning about money might begin by examining pennies, nickels, dimes, and quarters. Once they are familiar with these coin values, representing coins with bricks can help to further develop understanding. When learning about units of measurement, students can measure actual objects—such as toys, doors, and themselves—in both metric and customary units, before moving to more conceptual representations.

Research repeatedly demonstrates the need to develop meaning when teaching measurement (Van de Walle et al., 2012; Burns, 2007; Boaler, 2016). The best way to develop meaning for measurement is through physical models that help students to visualize measurement units. This book uses a representational approach to teach students basic measurement concepts, including: standard and non-standard units, length, time, money, customary liquid measures, metric measures, perimeter, and area.

HOW TO TEACH WITH THE BRICK MATH SERIES

Using the Teacher and Student Editions:
Start by taking students through the **Part 1: Show Them How** section of each chapter. Build the models, show them to the students, and ask students questions. Where directed, have students build the same models themselves so they are manipulating the bricks as you are guiding them. A document camera is helpful to display your models to the whole class as you build them. The step-by-step directions in the Teacher Edition work through several problems in Part 1. If you are using the companion Student Edition, have students draw their models and answer the questions in those books as you teach using the Teacher Edition.

Once students have mastered the modeling processes from Part 1, move to the **Part 2: Show What You Know** section of the chapter. Ask students to complete each of the problems using bricks and drawing their models. The companion Student Edition has space for writing answers and baseplate paper for drawing models. Move through the room and check that students are building their models correctly, drawing them clearly, and understanding the concepts being taught.

The Student Edition includes an assessment for each chapter, as well as additional problems for practice and challenge. The answer key for the chapter assessments can be found online on the Brick Math website, at www.brickmathseries.com/assessments. The book also includes an Assessment Chart to track each student's performance on all the skills taught in *Basic Measurement*.

Note: Active learning breeds active learners! Students will be motivated and engaged in math when they are using bricks. It will not be silent in your classroom, but it will be full of chatter about the math!

Suggested Bricks:

Brick Math is designed to be used with LEGO® bricks or LEGO® compatible bricks. If you already have bricks in your classroom, your students should be able to use them to make the models. They may have to combine smaller bricks together when the directions call for longer bricks such as 1x10s or 2x12s. Each student also needs a baseplate on which to build brick models.

Each chapter lists the bricks suggested for the lessons in that chapter for every two students, and the appendix includes a total brick inventory that lists all the bricks suggested for the program for every two students.

Specially designed Brick Math brick sets for one or two students are available for purchase from Brigantine Media. Brick sets are packaged in divided boxes and include a baseplate for each student.

Classroom Management Ideas:

- Before starting, have a conversation with the students about using bricks as a learning tool rather than a toy.
- Teach students the language of bricks (baseplate, stud, 1x1, 1x2, etc.).
- Assign brick sets to specific students and always give the same students the same sets. An easy way to do this is to number each brick set and assign the sets to pairs of students by number. When students know that they will always have to work with the same brick set, they are more likely to be careful that the bricks are returned to the set.
- Do not teach using bricks—or any manipulative—every day. Students also need to have opportunities to think through the math processes without having a physical object for modeling. Sometimes it helps to have students draw models without building them with bricks

first. Remember, they won't have access to manipulatives during most tests when they have to show what they have learned.

- To keep bricks clean: Put the bricks in a hosiery bag and wash them on the top rack of the dishwasher. Let them air dry. Clean bricks before assigning sets to new students.
- To keep bricks from sliding off desks, use foam shelf liner cut into rectangular pieces, or large meat trays (you can often get these free from a local supermarket).
- Inventory the sets twice a year and replace bricks as needed. There are a variety of vendors online that sell specific bricks, both new and used. LEGO® retail stores also sell a variety of individual bricks.

1

STANDARD UNITS AND COMPARING MEASURES OF LENGTH

SUGGESTED BRICKS

Size	Number
1x1	24
1x2	16
1x3	8
1x4	8
1x8	2
1x10	2

Note: Using a baseplate helps keep the bricks in place. Two large baseplates are suggested for these activities.

Students will learn/discover:

- How to compare measures using different units
- How to define standard unit of measure

Why is this important?

Students begin to study measurement by looking at various units of measure and learning that the standard unit is an important attribute of measurement. Using a standard unit of measure is important to precision in science, math, and engineering. Students should understand that using the same unit to measure different objects provides a frame of reference for comparison, which will lead to appropriate use of measurement concepts in more advanced math and real-world activities.

Vocabulary:

- **Attribute:** A quality or characteristic to be measured (e.g., length, distance, volume, time, etc.)
- **Unit of measure**: A quantity used to measure an attribute of a given object or event; units of measure may be *standard* or *nonstandard*
- **Standard unit of measure:** A quantifiable unit of measurement that is recognized by all members in a given area as part of a measurement system. For example, standard units of measure for length in the US include

feet, inches, yards, and miles, as defined by the customary measurement system. Standard units of measure for length in the UK include centimeters, meters, and kilometers, as defined by the metric measurement system.

- **Nonstandard unit of measure:** A unit of measurement that is not part of a recognized measurement system. Different nonstandard units of measure are not equivalent to one another (e.g., a given object measures differently in paper clips than pencils).

How to use the companion student book, *Basic Measurement Using LEGO® Bricks–Student Edition*:

- After students build their models, have them draw the models and explain their thinking in the Student Edition. Recording the models on paper after building them with bricks helps reinforce the concepts being taught.
- Discuss the vocabulary for each lesson with students as they work through the Student Edition.
- Use the assessment in the Student Edition to gauge student understanding of the content.

Part 1: Show Them How

Show students a 1x1 brick. Ask them what they notice about the 1x1 brick (*answer*: it has 1 stud).

Have students select another 1x1 brick and compare/contrast it with the first brick. Ask them how the bricks are alike and different (*possible answer:* both bricks have 1 stud but are different colors).

Explain to students that the color of the brick and the number of studs are considered *attributes* of the brick.

In this lesson, pairs of students will use different standard units to measure an assortment of everyday items. Before the lesson, collect a variety of small classroom objects for students to measure. Some suggestions are:

Unsharpened pencil	Marker
Jumbo paper clip	Crayon
Piece of notebook or copy paper	Straw
String	Ruler

1. Provide each pair of students with an unsharpened pencil. Ask students to select one brick of their choice. Explain to students that they will use bricks to measure the length of the pencil.

2. Place two baseplates side by side and put the pencil on the baseplates. Have students estimate how many of their chosen bricks are needed to equal the length of the pencil.

   Ask students to record their estimates before measuring, noting how many bricks of what size will equal the length of the pencil.

3. Have students measure the length of the pencil in bricks and record the measurement in bricks, noting how many bricks long the pencil measures and what size bricks were used.

   (*Answer for this example*: My pencil is six 1x4 bricks long.)

   Ask students to compare their answers to see which size of brick requires the most bricks to equal the length of the pencil.

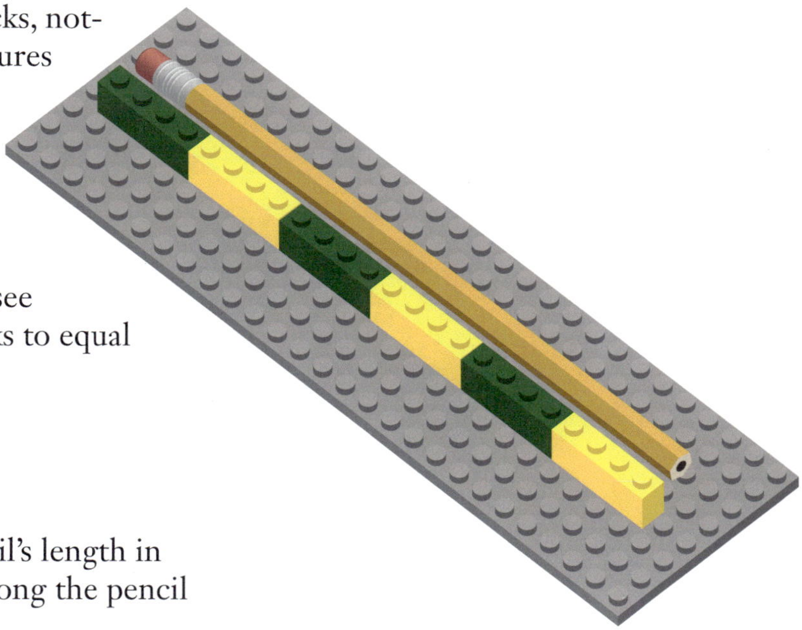

4. Ask students to count and record the pencil's length in number of studs, noting how many studs long the pencil measures.

   (*Answer for this example:* My pencil is 24 studs long.)

   Students should discover that all of the pencils are the same number of studs long, no matter which bricks are used as the unit of measurement. Students might also discover that they can use the idea of sets to find the number of studs. In this example, the pencil measures 6 bricks that each have 4 studs, or 6 sets of 4 (6 x 4). Depending on the level of the students, the concept of sets can be discussed.

5. Explain to students that in this case the stud serves as a standard unit of measure since it is always the same length. No matter which bricks are used, a stud measures the same amount every time. The bricks themselves are non-standard units of measure since each brick size is different. For example, it will take twelve 1x2 bricks to measure the pencil, or eight 1x3 bricks.

6. Have students repeat this process with another item (see suggested list) and discuss what they discover.

7. Have students place a ruler next to the bricks and the pencil. *Note*: Be sure to discuss the parts of a ruler and what each line means on the ruler prior to starting this part of the activity.

   Show students how to start measuring at the end of the ruler. Direct students to use either the centimeter or the inches side, depending on their age and your choice.

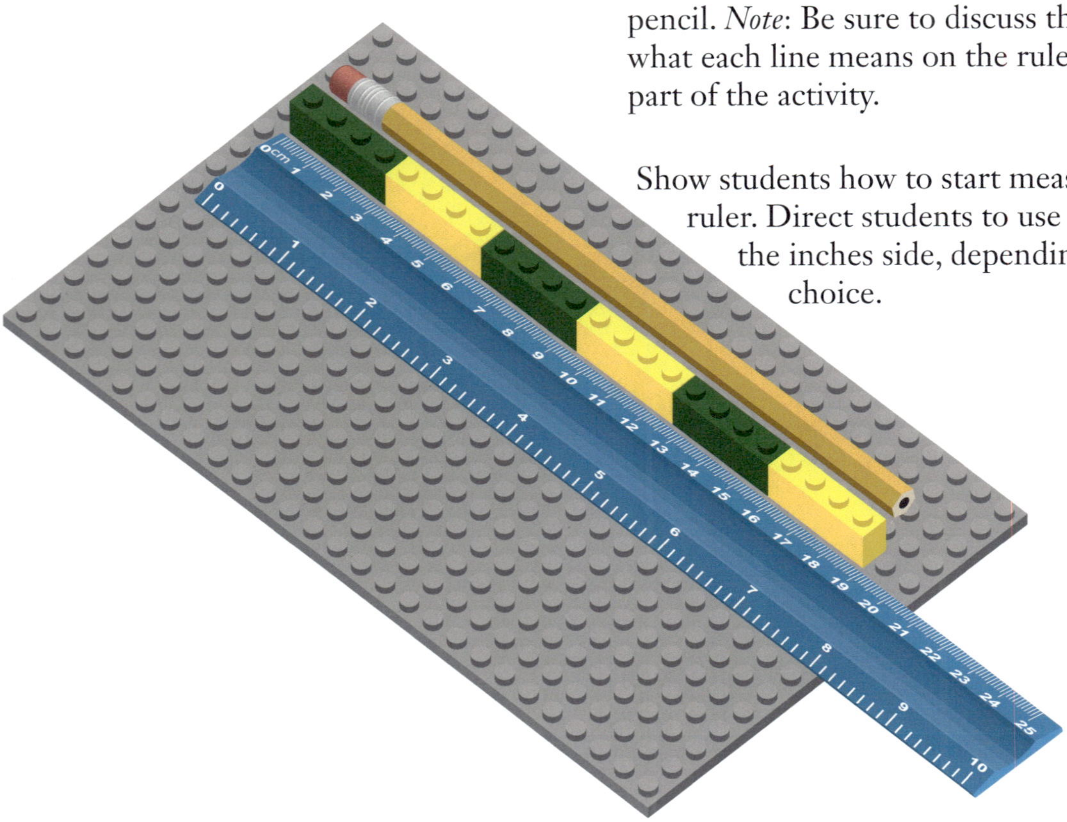

8. Ask students to observe where the pencil ends and record its length in number of centimeters or inches. In this example, the pencil is about 19 centimeters long (or 7½ inches long). This comparison shows students another standard measurement tool.

9. Repeat this with other items you have collected. Have students record and make statements.

Part 2: Show What You Know

Provide each pair of students with an assortment of at least four items to measure in bricks and studs. Choose any items (see suggested list), but make sure to include items that you have not already measured in Part 1. Have them experiment with at least three sizes of bricks to make sure they understand the idea of standard stud measures.

1. Measure each item using the 1x1 brick and three additional brick sizes of your choice. Find the length of the item in number of bricks and number of studs. What did you discover?

   Note: Provide or have students create a chart for their data collection. Answers will vary depending on the type and size of the materials.

   Sample chart:

Brick	Paper clip		String		Marker		Straw	
	Bricks	Studs	Bricks	Studs	Bricks	Studs	Bricks	Studs
1x1								
1x2								
1x3								
1x4								

2. Using the bricks and your ruler, measure and record the length of five bricks in centimeters and/or inches.

Note: Provide or have students create a chart for their data collection. Have students answer this question only if they have been introduced to the ruler and know what the marks on the ruler represent.

Sample chart:

Brick	Centimeters	Inches
1x2	$1\frac{1}{2}$ cm	$\frac{9}{16}$ in
1x4	3 cm	$1\frac{1}{4}$ in
1x6	$4\frac{7}{10}$ cm	$1\frac{14}{16}$ in
1x8	$6\frac{2}{10}$ cm	$2\frac{1}{2}$ in
1x10	8 cm	$1\frac{1}{16}$ in

2

MONEY

SUGGESTED BRICKS

Size	Number
1x1	Up to 100
1x2	10
1x3	8
1x4	8
1x10	4

Note: Using a baseplate helps keep the bricks in place. One baseplate is suggested for these activities.

Students will learn/discover:

- How to model coin values up to one dollar, using bricks
- How to add coin values (pennies, nickels, dimes, and quarters) within one dollar

Why is this important?

Understanding monetary amounts is a real-world skill that students must master in order to purchase material goods and provide for their needs in life. Students should be able to identify the amounts that total up to one dollar, understand how to calculate amounts necessary to purchase items, and understand how to determine the amount left over after a purchase. These skills require a firm knowledge of monetary units.

Vocabulary:

- Penny: A monetary value equivalent to 0.01 of a dollar; the smallest unit of US coinage
- Nickel: A monetary value equivalent to 0.05 of a dollar
- Dime: A monetary amount equivalent to 0.10 of a dollar
- Quarter: A monetary amount equivalent to 0.25 or ¼ of a dollar
- Dollar: A monetary amount equivalent to 100 pennies, 20 nickels, 10 dimes, or 4 quarters
- Add: To join or put together
- Cents sign: ¢
- Dollar sign: $

How to use the companion student book, *Basic Measurement Using LEGO® Bricks–Student Edition*:

- After students build their models, have them draw the models and explain their thinking in the Student Edition. Recording the models on paper after building them with bricks helps reinforce the concepts being taught.
- Discuss the vocabulary for each lesson with students as they work through the Student Edition.
- Use the assessment in the Student Edition to gauge student understanding of the content.

Part 1: Show Them How

Explain to students that they will learn to model monetary values using bricks.

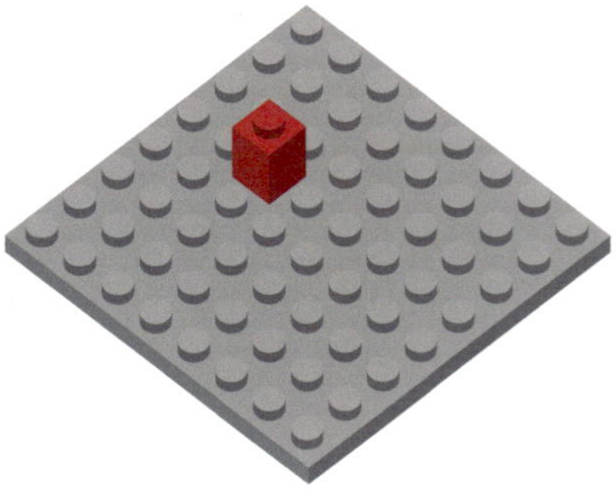

1. Model 1 penny with one 1x1 brick. Show students how to write 1 cent ($0.01 or 1¢).

Have students build the same model, then draw and label their models.

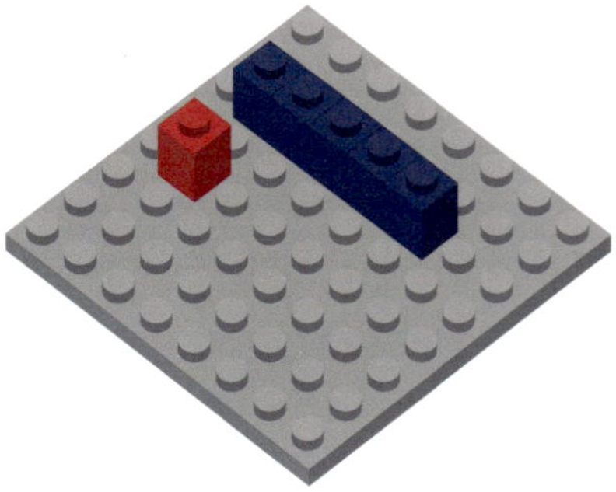

2. Ask students how many 1x1 bricks are needed to model a nickel. Have them build a model of a nickel using five 1x1 bricks of the same color. Ask students to write 5 cents ($.05 or 5¢), then draw and label their models.

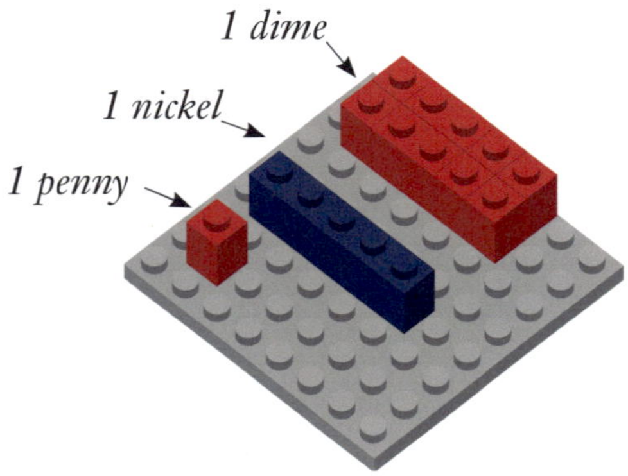

3. Ask students how to model a dime (*possible answers*: ten 1x1 bricks of the same color or one 1x10 brick).

Note: Make sure students understand that they can also model a dime with two sets of five 1x1 bricks (the same as ten 1x1 bricks, but in sets). This will help them begin to use sets as part of a dollar when counting money. Students learn to count by fives and tens to 100, so the idea of building sets of 5 is an extension of previously learned math content as well as preparatory for multiplication.

Have students build two sets of five 1x1 bricks, then draw and label their models.

4. Ask students to name the coin values modeled (*answer:* one penny, one nickel, and one dime). Have students identify the total amount of money modeled on the baseplate (*answer:* 16 cents). Ask students to describe how they determined the total amount (*possible answer:* count by fives and add 1). Have students draw and explain their models.

5. Have students build five sets of nickels.

Have students count by fives and identify the value of the coins modeled (*answer*: 25 cents). Have students record the monetary amount of 5 nickels in correct syntax (*answer*: $.25 and 25¢). *Note:* If students struggle with counting by fives, they can count each stud using one-to-one correspondence. Remind students that one stud is equivalent to 1 cent.

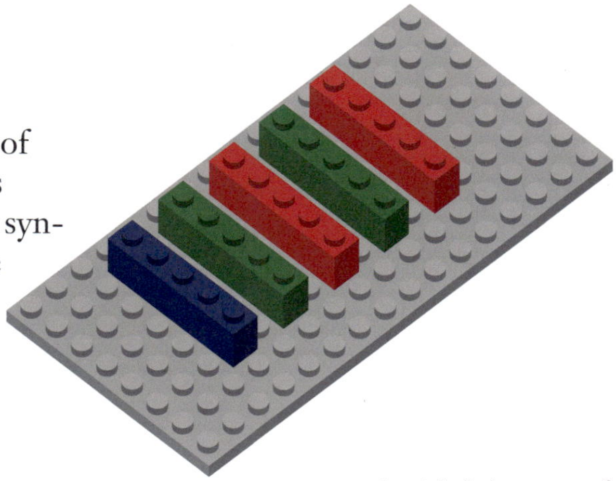

Looking at this model, lead students to discover the number of dimes in 25 cents (*answer:* 2 dimes, with 1 nickel left over). *Note:* To discover this, students might group sets of 5 pennies to make two sets of 10 pennies, as illustrated; they will then realize that one set of 5 pennies (one nickel) remains.

Have students draw their models and label the amounts.

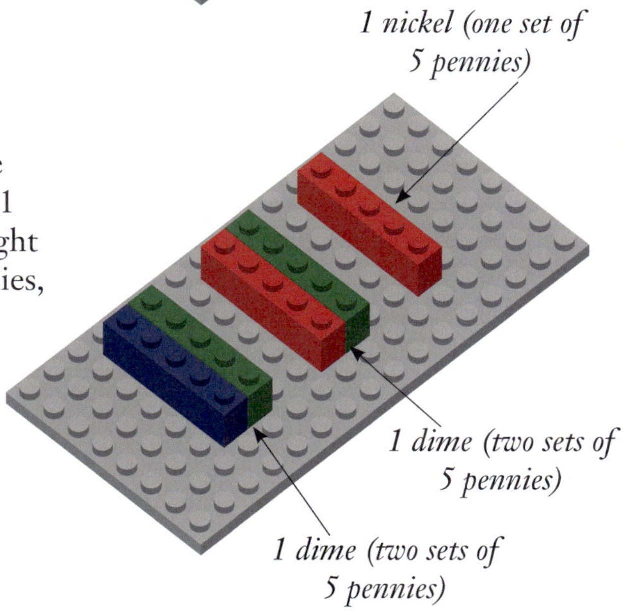

6. Challenge students to work in pairs to create a dollar. Pose the following question: Can you and your partner find a way to model one dollar?

Let students use bricks to model one dollar using pennies, nickels, or dimes. Discuss possibilities with the class after they have created some models.

Ask students how many pennies are in a dollar (*answer:* 100 pennies), how many nickels are in a dollar (*answer:* 20 nickels), and how many dimes are in a dollar (*answer:* 10 dimes).

Note: If students have learned how to model place value with bricks, you can review this concept now. Students can exchange ten 1x1 bricks for one 1x2 brick, and ten 1x2 bricks for one 1x3 brick to show the tens and hundreds place values (See *Addition Using LEGO® Bricks—Teacher Edition*).

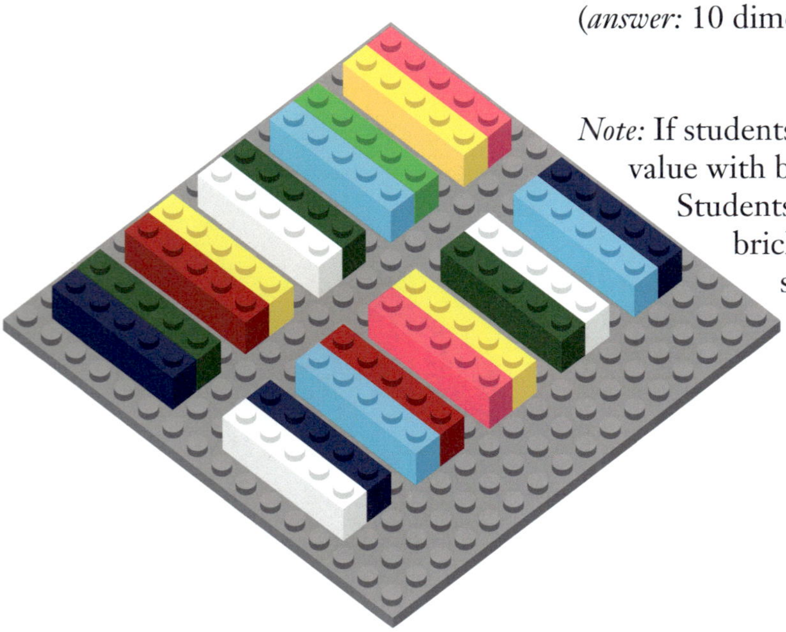

Possible solution:
Ten sets of 10 pennies, or ten dimes; or twenty sets of 5 pennies, or twenty nickels; or 100 pennies

Part 2: Show What You Know

1. Can you build a model that shows $0.55? Draw your model and label each coin value in your model. How many nickels are in your model? How many dimes are in your model? Explain your thinking.

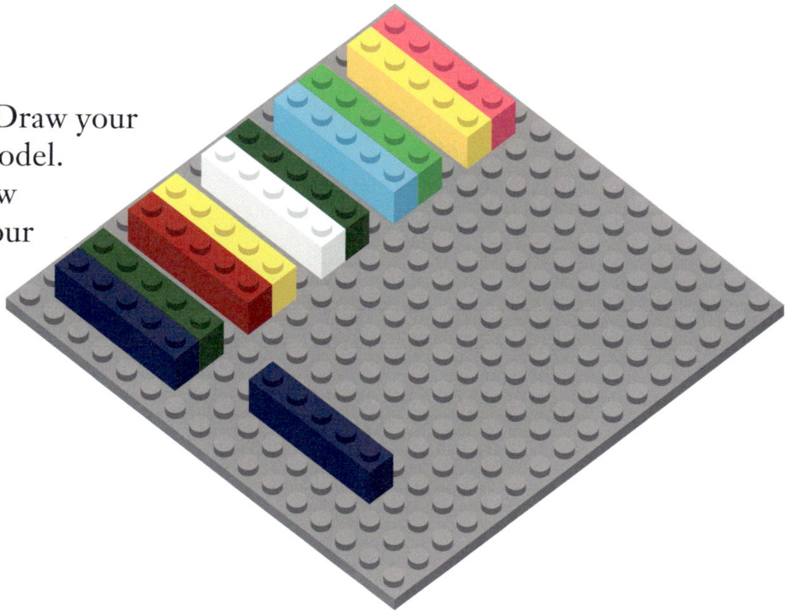

Possible solution:
5 sets of 10 pennies and 1 set of 5 pennies; or 5 dimes and 1 nickel; or 11 sets of 5 pennies; or 11 nickels

2. Can you build a model that shows $0.75? Draw your model and label each coin value in your model. How many nickels are in your model? How many dimes are in your model? How many quarters are in your model? Explain your thinking.

Possible solutions:

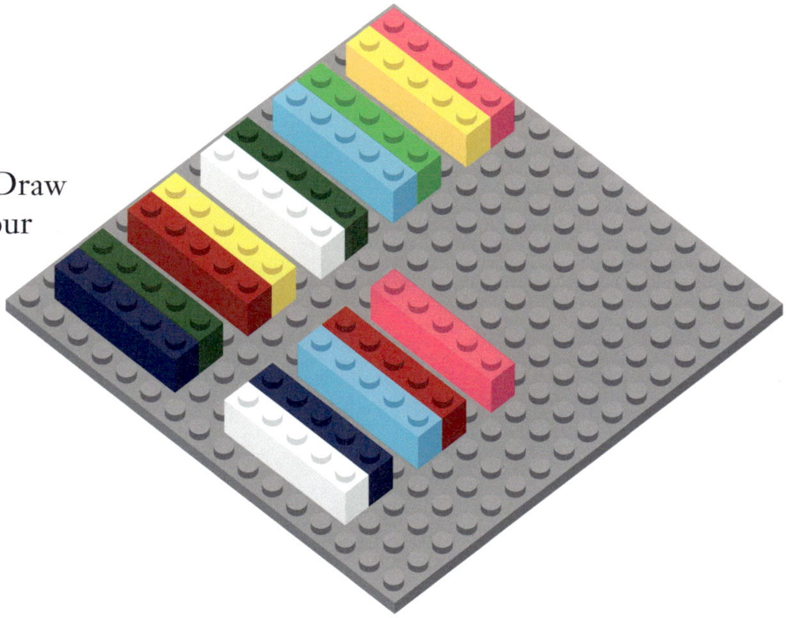

15 nickels; or 15 sets of 5 pennies: or 7 dimes and 1 nickel; or 7 sets of 10 pennies and 1 set of 5 pennies; or 75 pennies

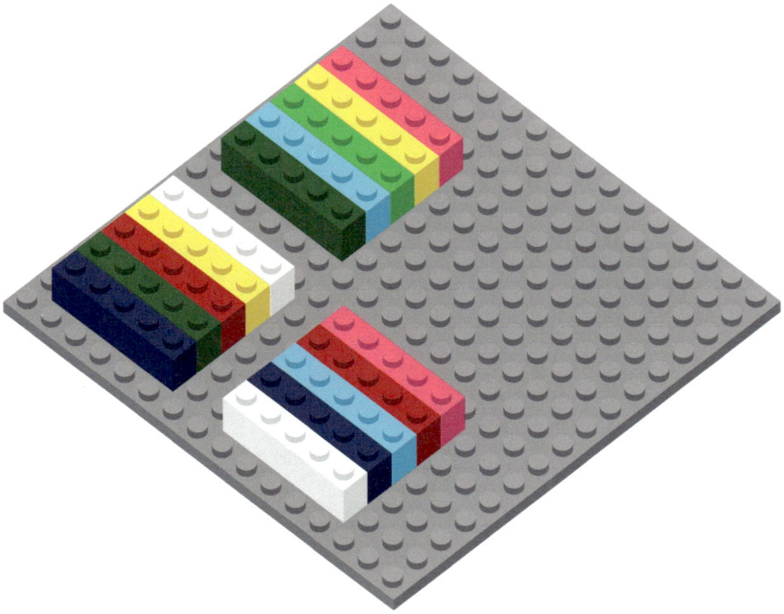

3 quarters; or
3 sets of 25 pennies

3. Can you build a model that shows two ways to make $0.60? Draw your model and explain your thinking.

Possible solutions:

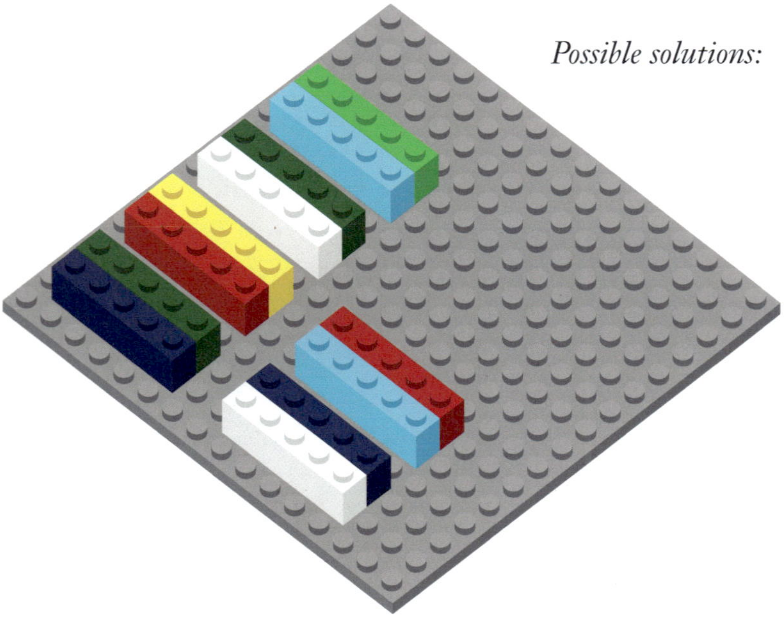

6 dimes; or 6 sets of 10 pennies

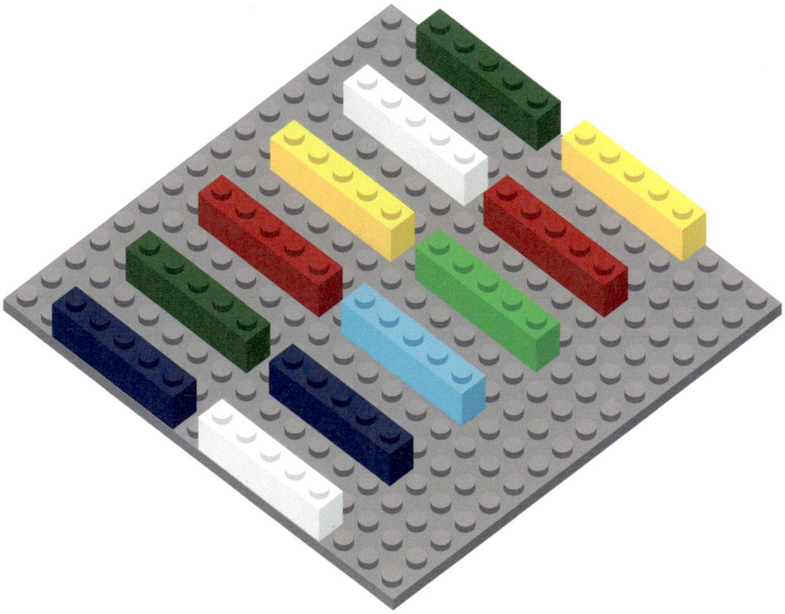

12 nickels; or 12 sets of 5 pennies

4. Partner activity: Choose an amount less than one dollar, but do not tell your amount to your partner. Build a model of your amount. Show your model to your partner. Tell your partner the total number of nickels, dimes, or quarters in model by describing your model in this way:

"My model has _____ (number) quarters, ____ (number) dimes, _____ (number) nickels, and ____ (number) pennies."

Have your partner identify the amount, then switch roles so you each have a turn to create a model.

Models will vary.

3

SUGGESTED BRICKS

Size	Number
1x1	10 (including 4 red bricks)
1x2	6
1x3	2
1x4	2
1x6	2
1x8	2
1x10	2
1x12	2

Note: Using a baseplate helps keep the bricks in place. Five 5″ x 5″ baseplates or three 10″ x 10″ baseplates are suggested for these activities. Students can work in groups to join baseplates when creating the linear clock with bricks.

INTRODUCTION TO TIME

Students will learn/discover:
How to tell time to the nearest hour, half hour, and 5-minute intervals

Why is this important?
The ability to tell time is a basic and necessary skill, foundational to understanding elapsed time (how much time has passed), sequencing events, and being on time for events in daily life. Students should be able to determine what time it is, given a start time and the number of hours and minutes that have elapsed since then, to the nearest hour, half hour, and 5-minute increment. Because parts and types of clocks cannot be taught with bricks, it is important that teachers also discuss actual clocks with students and explain the hands on a regular analog clock.

Vocabulary:

- Analog clock: Clock with a face featuring numbers 1-12, which represent both the hour and the minutes in 5-minute increments
- Digital clock: Clock that uses only numbers to display hours, minutes, and sometimes seconds
- Hour hand: The shorter hand on an analog clock, which points to the hour
- Minute hand: The longer hand on an analog clock, which points to the minute

How to use the companion student book, *Basic Measurement Using LEGO® Bricks–Student Edition*:

- After students build their models, have them draw the models and explain their thinking in the Student Edition. Recording the models on paper after building them with bricks helps reinforce the concepts being taught.
- Discuss the vocabulary for each lesson with students as they work through the Student Edition.
- Use the assessment in the Student Edition to gauge student understanding of the content.

Part 1: Show Them How

Discuss analog and digital clocks with students, and explain the meaning of the hands on an analog clock.

Explain to students that they will build a linear clock, using it much like a timeline to learn how to use skips (also known as jumps) to tell time to the nearest hour, half hour, and 5-minute increments.

Build a linear time clock with students as illustrated. *Note:* For the sake of time and consistency, you may want to build this for the students before the lesson, rather than having students build it.

> **Step 1:** Place five baseplates side by side horizontally. Join the baseplates together at the top with 1x2 bricks, but make sure students don't confuse these with the bricks that are part of the clock when they are counting to tell time.
>
> **Step 2:** Place a 1x12 brick flush left on the leftmost baseplate. Explain to students that this brick represents the "12" on the clock.
>
> **Step 3:** The top row of this 1x12 brick is now the "top" of the clock. Moving horizontally to the right in this row, skip 11 studs on the baseplate. Place a (non-red) 1x1 brick on the baseplate at the twelfth stud. This brick represents the "1" on the clock.

Step 4: Moving to the right of the 1x1 brick, skip 11 more studs on the baseplate. At the twelfth stud, place a 1x2 brick vertically on the baseplate. *Note:* The top of the 1x2 brick should be flush with the top of the 1x1 and 1x12 bricks. This brick represents the "2" on the clock.

Step 5: Moving to the right of the 1x2 brick, skip 11 studs on the baseplate. At the twelfth stud, place a 1x3 brick vertically on the baseplate to represent the "3" on the clock.

Step 6: Moving to the right of the 1x3 brick, skip 11 studs on the baseplate. At the twelfth stud, place a 1x4 brick vertically on the baseplate to represent the "4" on the clock.

Step 7: Moving to the right of the 1x4 brick, skip 11 studs on the baseplate. At the twelfth stud, place one 1x4 and one 1x1 of the same color together vertically on the baseplate to represent the "5" on the clock.

Step 8: Moving to the right, skip 11 studs and place a 1x6 brick vertically on the baseplate to represent the "6" on the clock.

Step 9: At this point, you will no longer have room to continue building your clock in a straight line. To skip 11 studs, begin by counting the number of studs left in the row. Then continue counting on the bottom row of the baseplate, beginning flush left as seen in the illustration. Place one 1x6 and one 1x1 brick of the same color together vertically on the baseplate to represent the "7" on the clock.

Steps 10–12: Continue to add bricks vertically starting at the bottom row of the baseplate, skipping 11 studs each time (add a 1x8 brick to represent "8," a 1x1 and a 1x8 brick to represent "9," a 1x10 brick to represent "10," a 1x1 and a 1x10 brick to represent "11," and a 1x12 brick to represent "12").

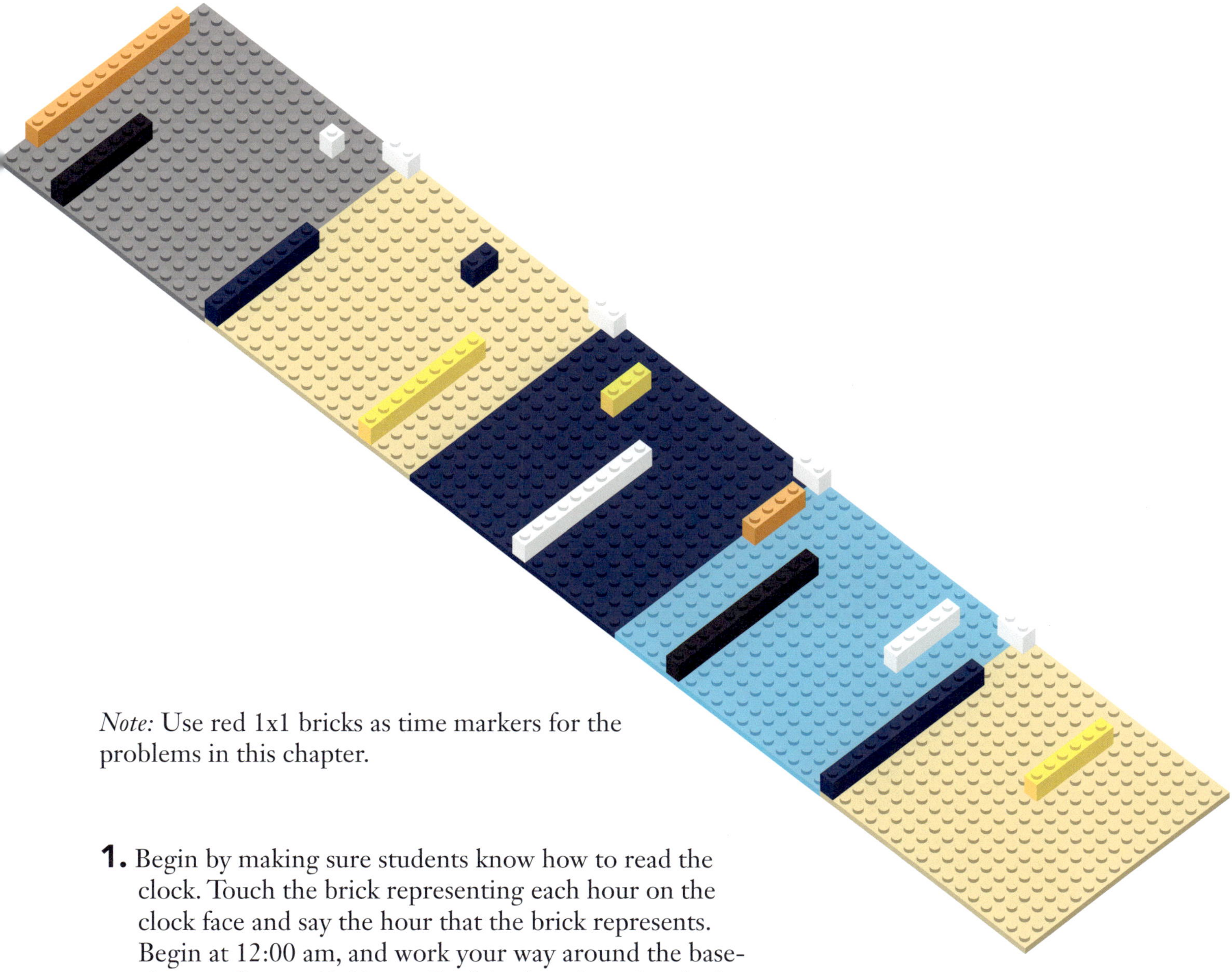

Note: Use red 1x1 bricks as time markers for the problems in this chapter.

1. Begin by making sure students know how to read the clock. Touch the brick representing each hour on the clock face and say the hour that the brick represents. Begin at 12:00 am, and work your way around the baseplate, ending at 12:00 pm. Explain that since the clock models 12 hours, or only half the day, it can be read as either am or pm.

Explain to students that the distance between side-by-side studs on the baseplate represents a 5-minute interval. Practice counting by 5-minute intervals, 10-minute intervals, and half-hour intervals by touching the studs that represent each time.

2. Place a red 1x1 brick on the stud halfway between 12:00 and 1:00. Ask students what time the red brick represents (*answer:* 12:30 am).

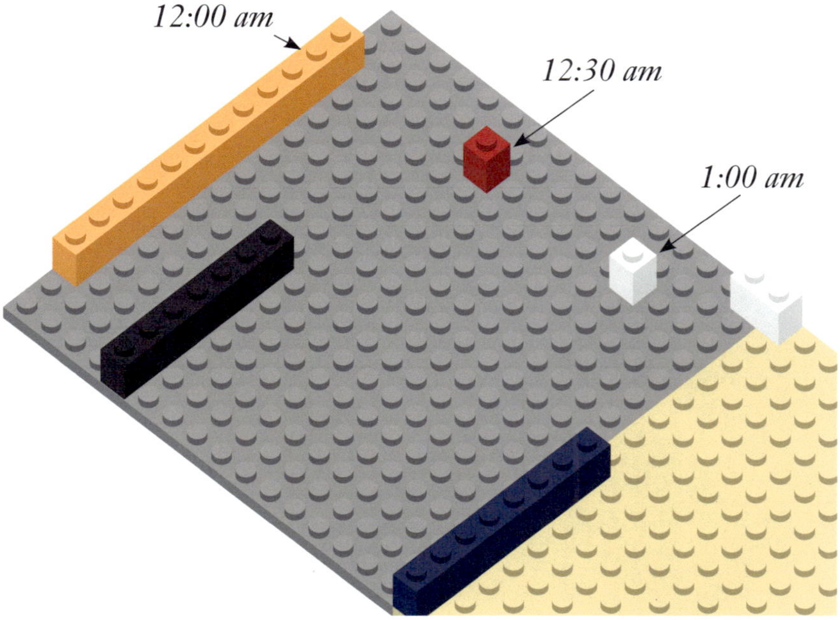

Ask students what time it will be in 2 hours (*answer:* 2:30 am). Have them place the second red 1x1 brick on the stud representing 2:30 am.

Explain to students how to combine hours and minutes in a math sentence that shows this time problem (*answer:* 12 hours 30 minutes + 2 hours = 12 hours + 2 hours + 30 minutes or 2:30 am).

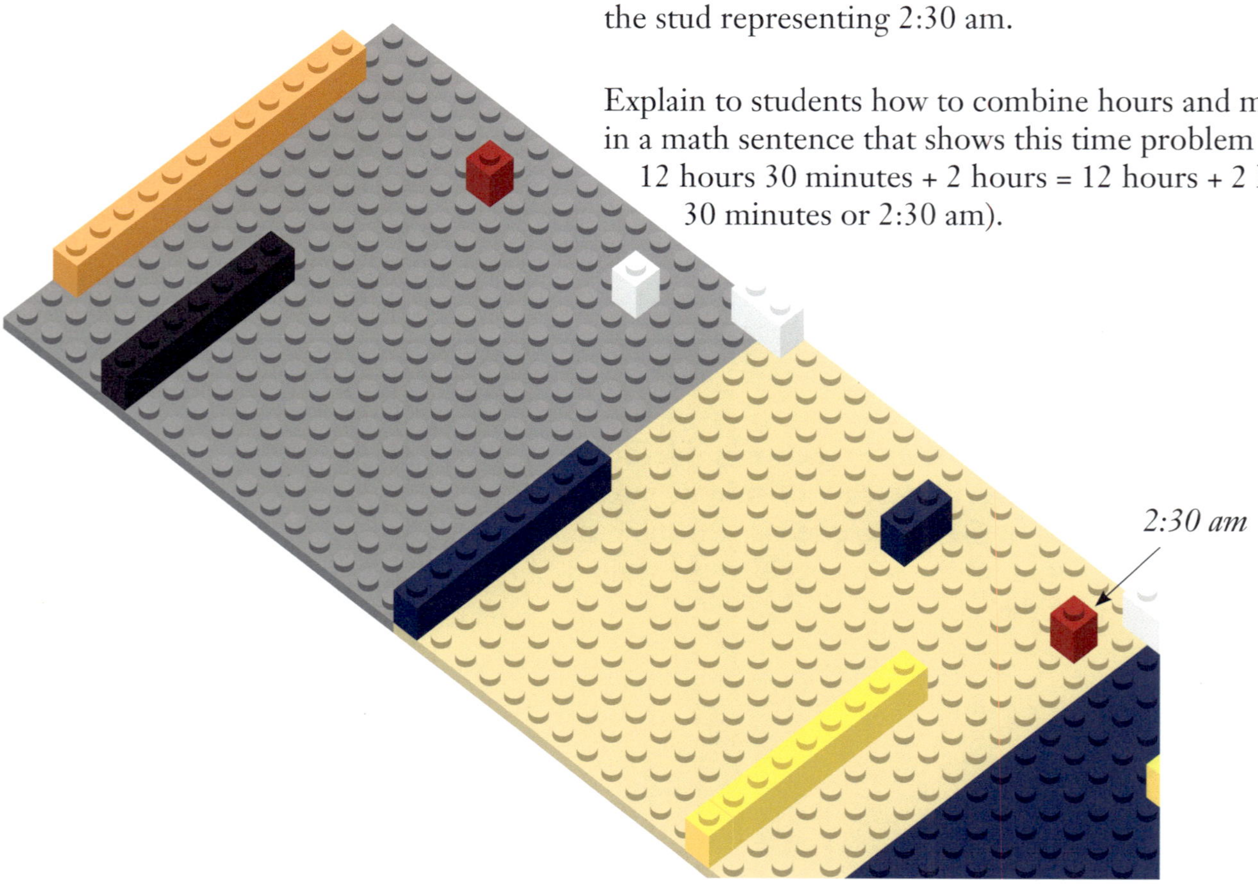

3. Tell students that the time is now 3:30 am. Place a red 1x1 brick on the stud that represents this time. Ask students to determine what time it will be in 15 minutes. Show the new time by placing a second red brick on the corresponding stud (*answer:* 3:45 am). Ask students to write a math sentence showing their model (*answer:* 3 hours 30 minutes + 15 minutes = 3 hours 45 minutes or 3:45 am)

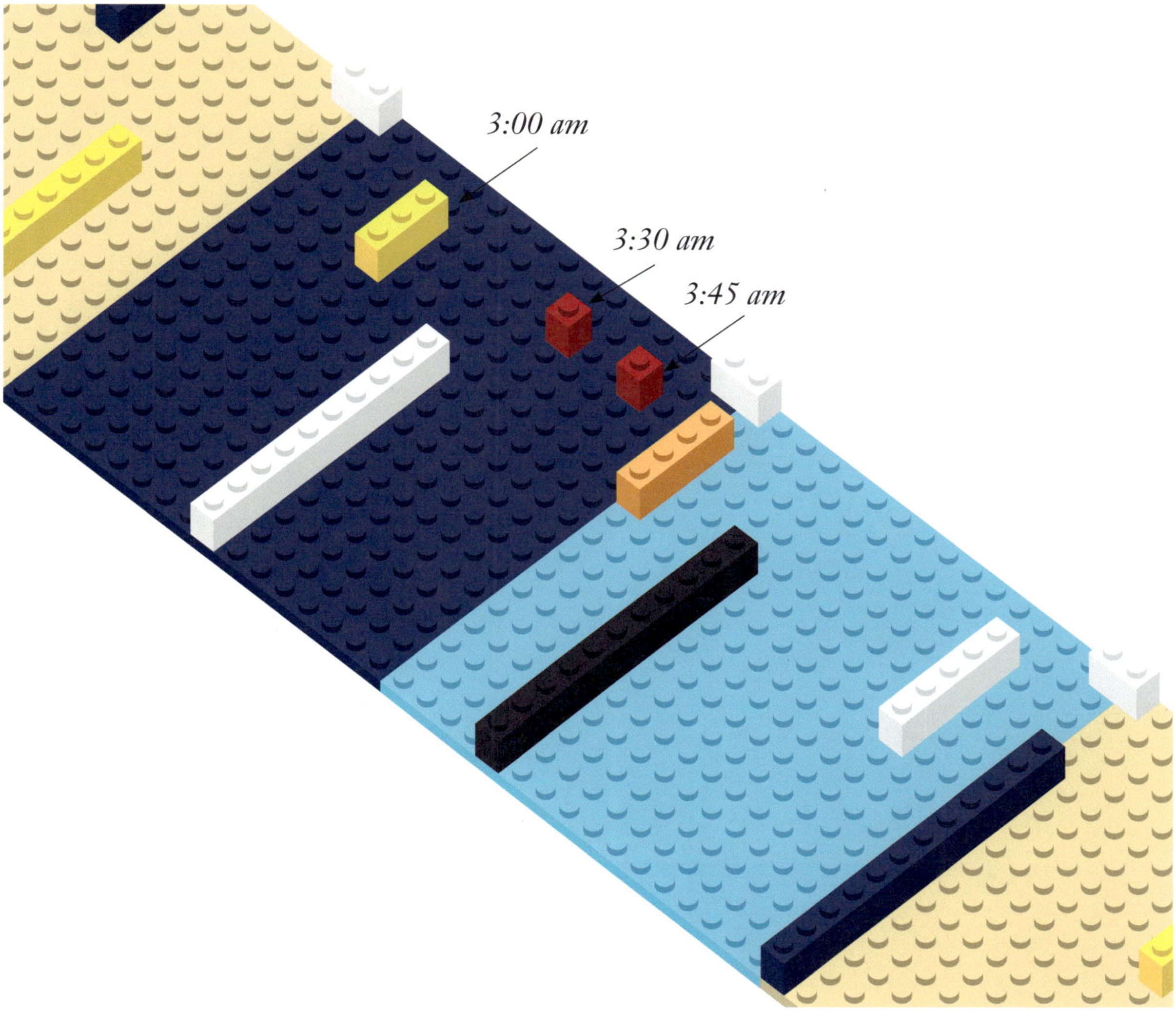

4. Tell students the time is now 6:15 am. Tell students that you want to determine what time it will be in 1 hour and 55 minutes.

Have students place a red 1x1 brick on the baseplate at 6:15 am. Ask students to count forward 1 hour and identify the time (*answer:* 7:15 am). Then, ask students to count forward another 45 minutes, place another red 1x1 brick on the baseplate as a benchmark, and identify the time (*answer:* 8:00 am). Finally, ask students to add 10 minutes by moving forward on the clock and placing another red 1x1 brick on the baseplate. The final time is 8:10 am.

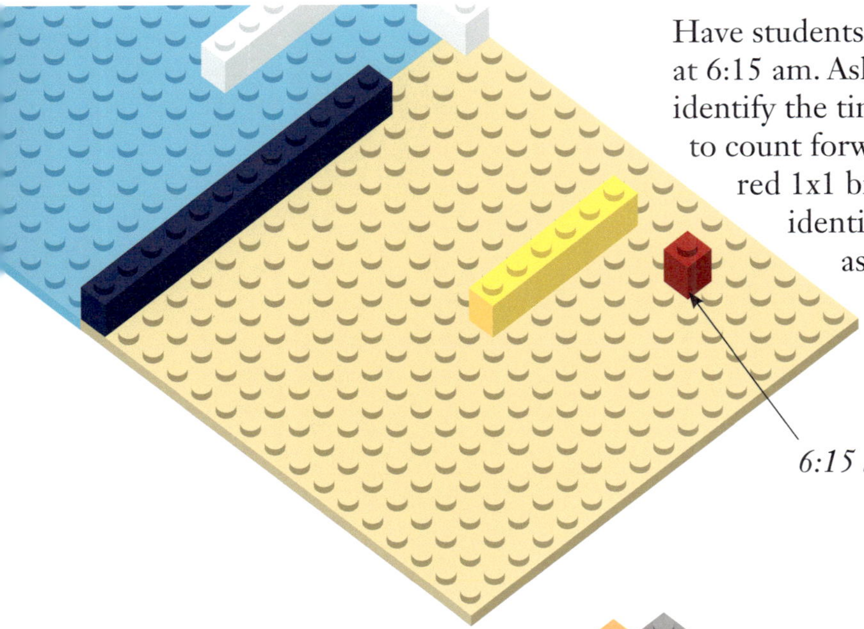

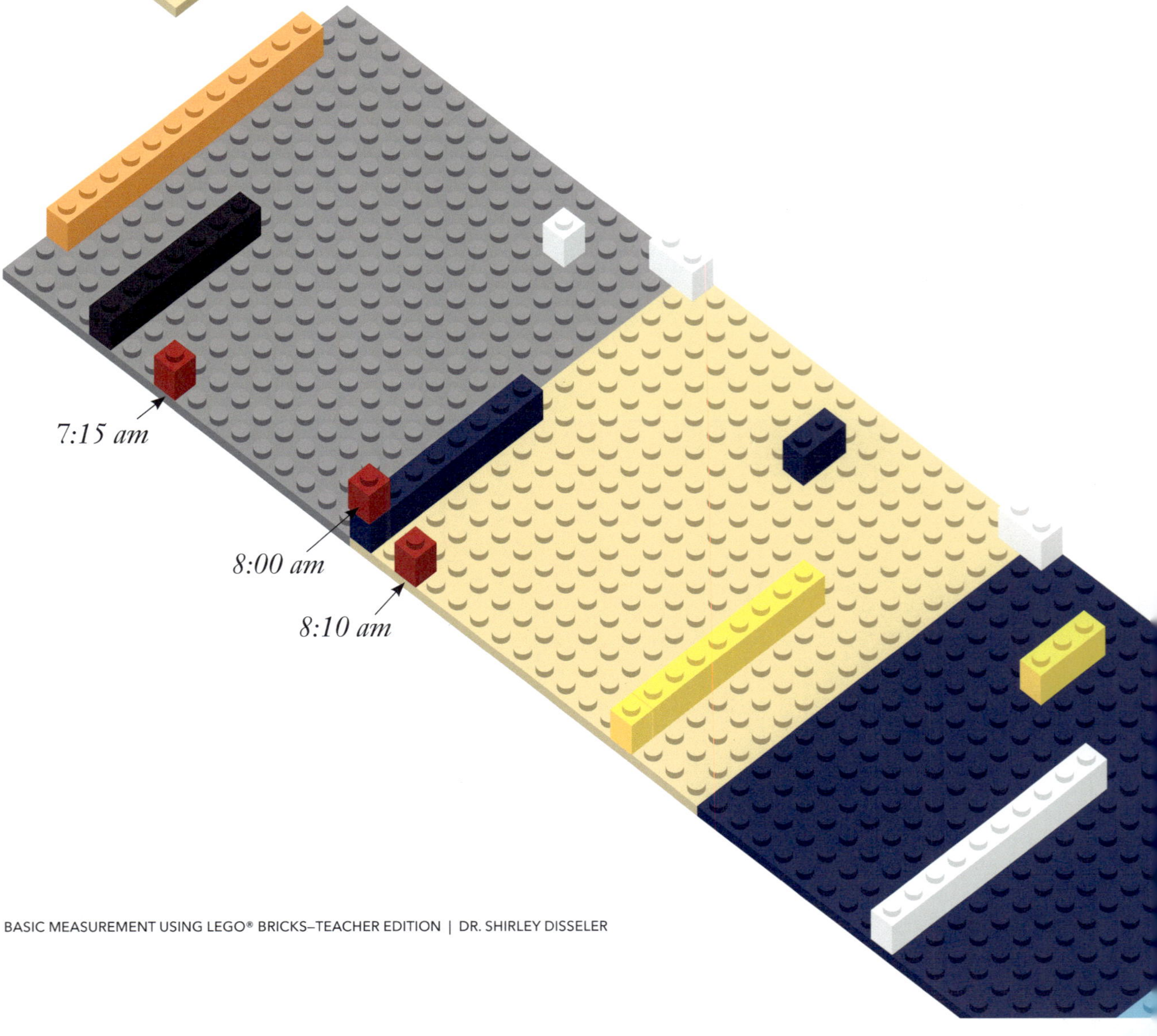

Part 2: Show What You Know

1. Using the linear clock model, place a red 1x1 brick at 2:00 pm on the baseplate. Count forward 65 minutes, and mark the new time on the baseplate with another red 1x1 brick. What time is it now? Draw the portion of the clock that shows the elapsed time. Write a math sentence for your model.

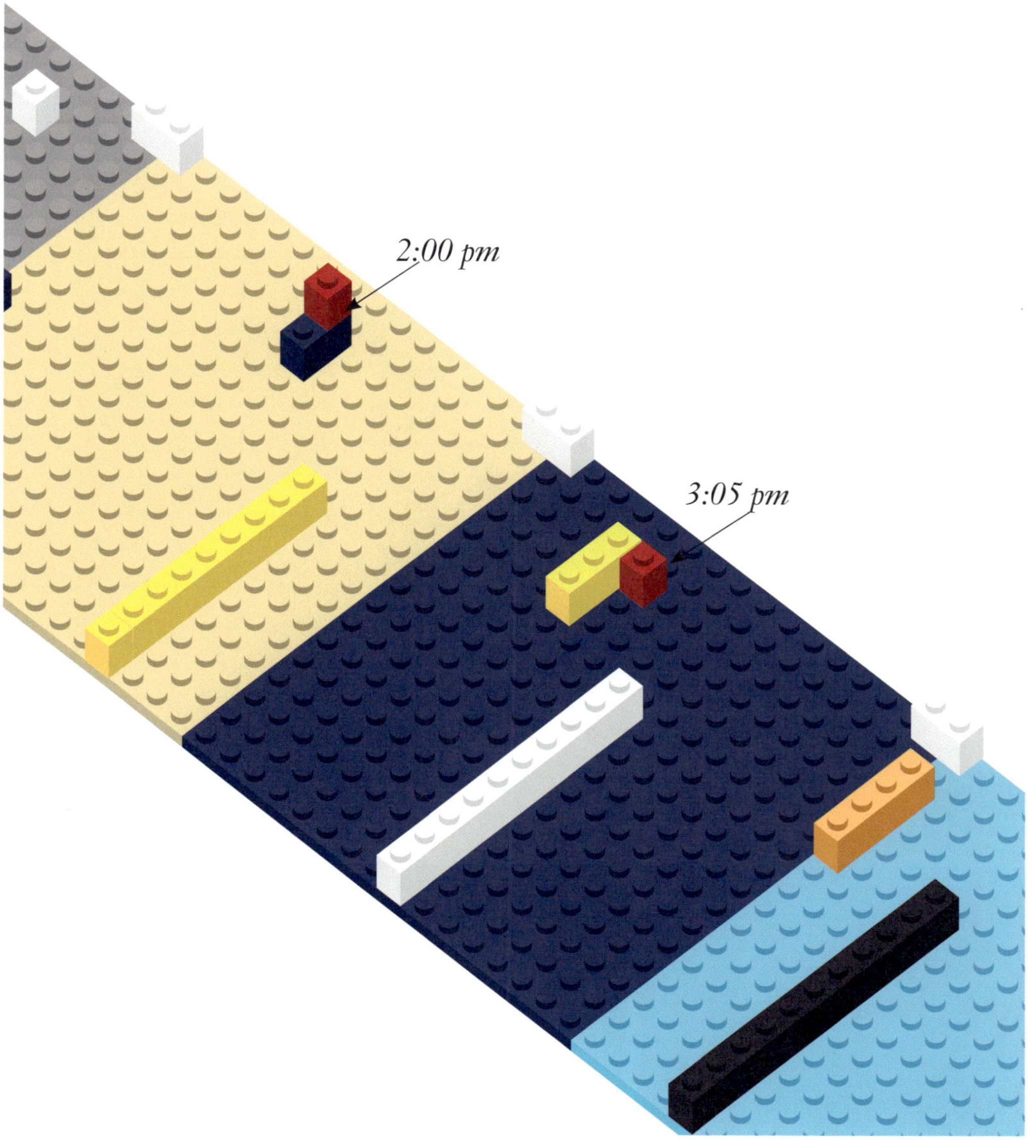

2 hours + 65 minutes =
3 hours 5 minutes or 3:05 pm

2. Using the linear clock model, start at the red 1x1 brick on the baseplate at 3:05 pm. Count forward 25 minutes and mark the new time on the baseplate with another red 1x1 brick. What time does it show? Draw the portion of the clock that shows the elapsed time. Write a math sentence for your model.

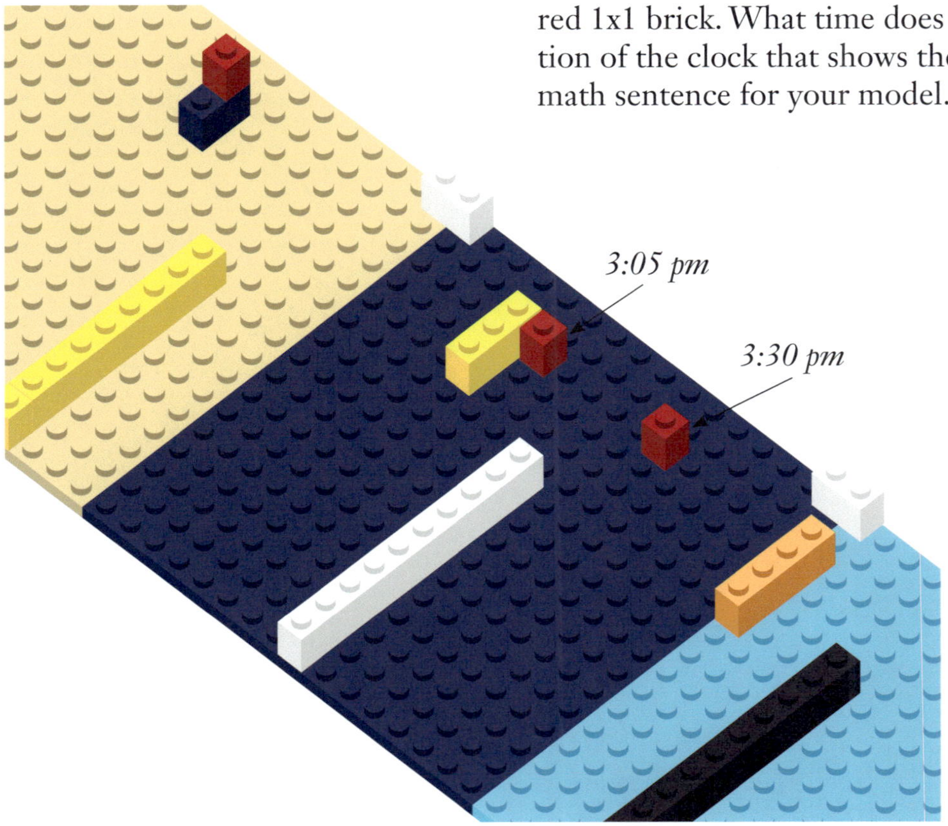

3 hours 5 minutes + 25 minutes =
3 hours 30 minutes or 3:30 pm

3. If it is 6:30 pm now, what time will it be in 4 hours and 15 minutes? Build a model to show the new time. Draw your model, showing how you added hours first, then minutes. Explain your thinking.

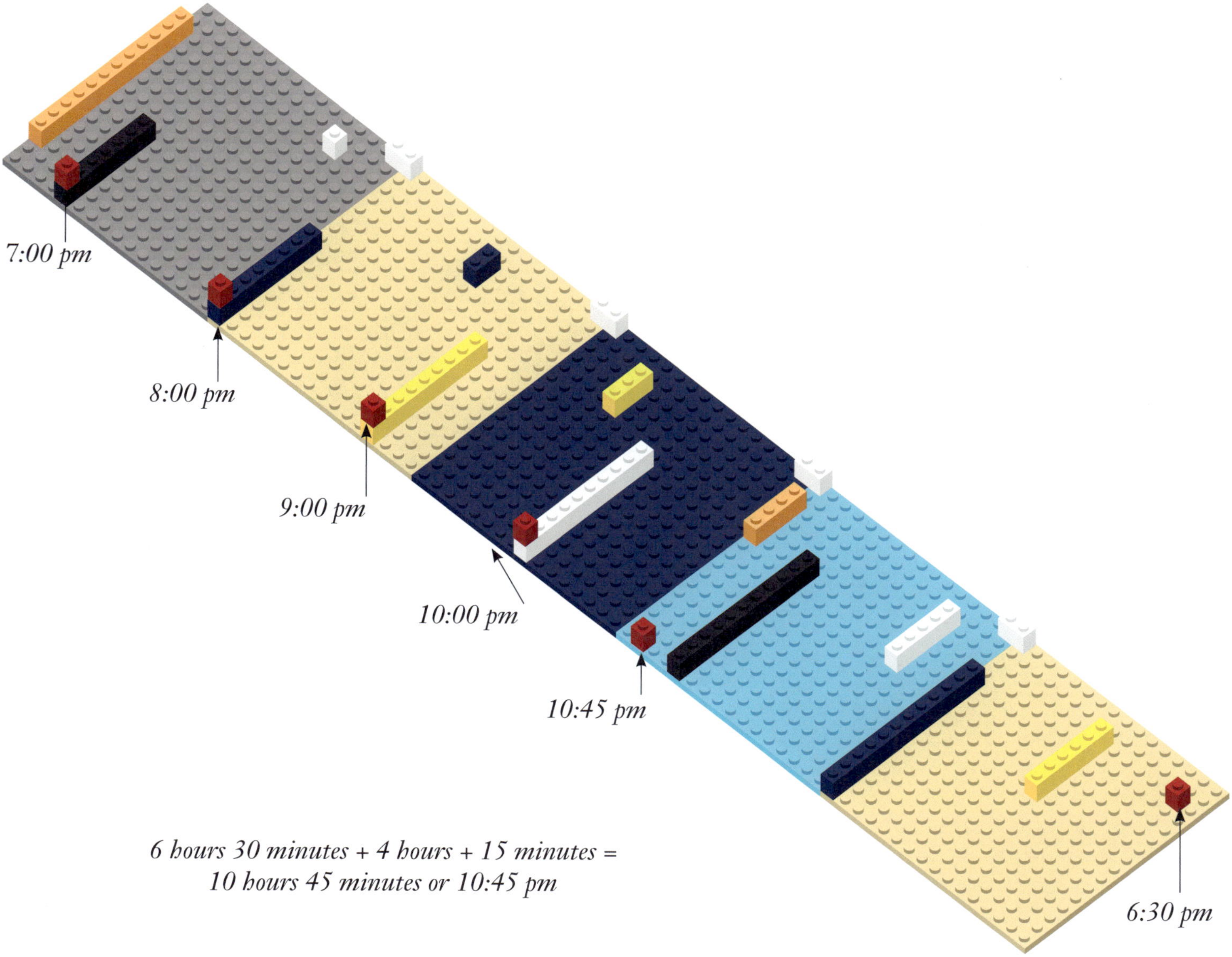

6 hours 30 minutes + 4 hours + 15 minutes =
10 hours 45 minutes or 10:45 pm

4. Place a red brick at 4:15 pm on the clock. Count backward 2 hours and 5 minutes, and mark the new time with another red brick. What time is it now? Draw your model and explain your thinking.

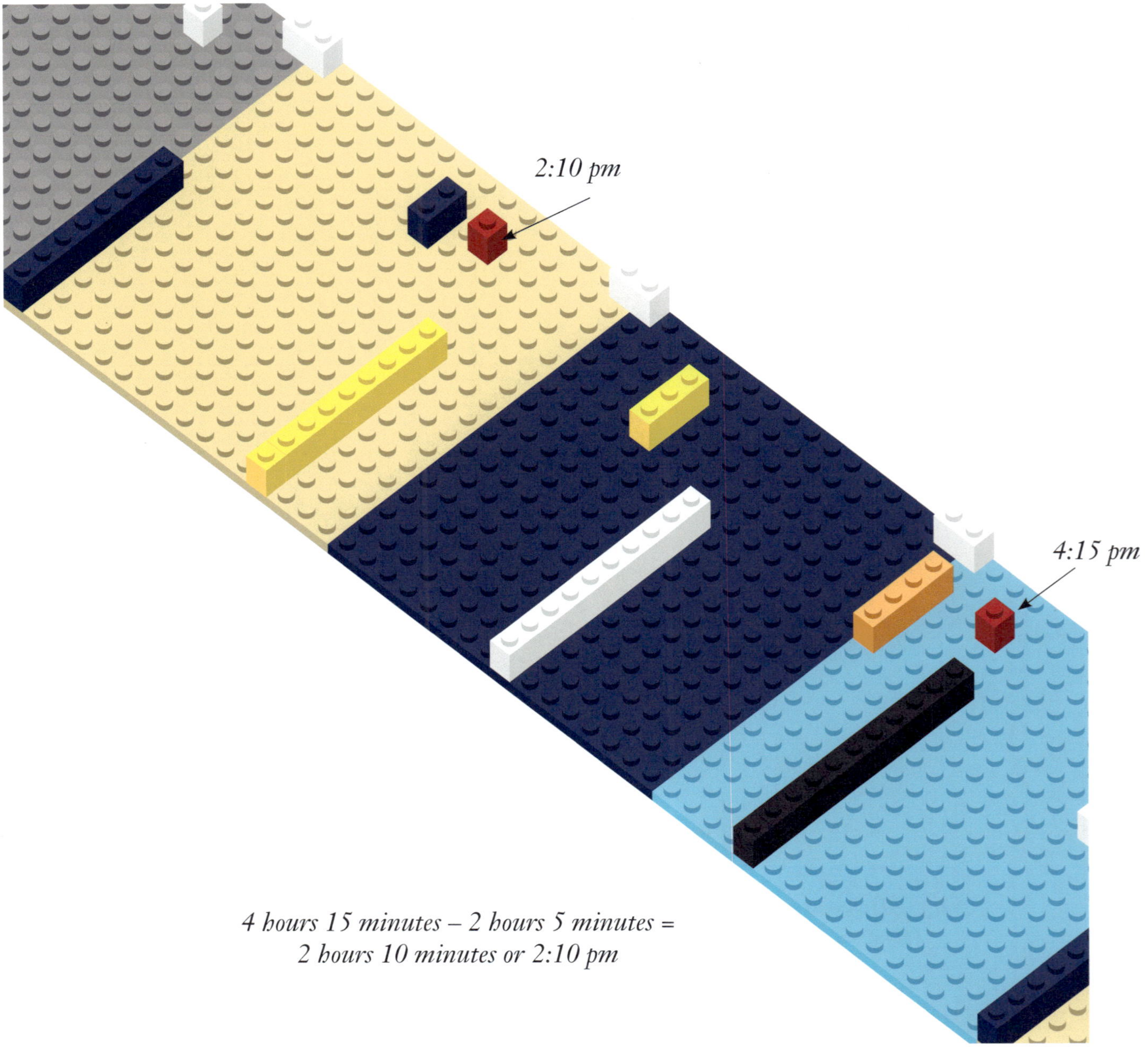

4 hours 15 minutes – 2 hours 5 minutes = 2 hours 10 minutes or 2:10 pm

4

ELAPSED TIME

Students will learn/discover:
How to determine elapsed time (how much time has passed between events), using their knowledge of telling time and reading the clock

Why is this important?
Understanding elapsed time and sequencing helps people manage their time and create a daily schedule. In everyday life, determining elapsed time is key to activities such as: arriving on time for appointments, baking, taking medications, and planning trips. Because parts and types of clocks cannot be taught with bricks, it is important that teachers also discuss actual clocks with students and explain the hands on an analog clock.

Vocabulary:

- Analog clock: Clock with a face featuring numbers 1-12, which represent both the hour and the minutes in 5-minute increments
- Digital clock: Clock that uses only numbers to display hours, minutes, and sometimes seconds
- Elapsed time: The time that passes between one event and another
- Hour hand: The shorter hand on an analog clock, which points to the hour
- Minute hand: The longer hand on an analog clock, which points to the minute

SUGGESTED BRICKS

Size	Number
1x1	9 (including 4 red bricks)
1x2	4
1x3	2
1x4	2
1x6	2
1x8	2
1x10	2
1x12	2

Note: Using a baseplate helps keep the bricks in place. Five 5″ x 5″ baseplates or three 10″ x 10″ baseplates are suggested for these activities. Students can work in groups to join baseplates when creating the linear clock with bricks.

How to use the companion student book, *Basic Measurement Using LEGO® Bricks–Student Edition*:

- After students build their models, have them draw the models and explain their thinking in the Student Edition. Recording the models on paper after building them with bricks helps reinforce the concepts being taught.
- Discuss the vocabulary for each lesson with students as they work through the Student Edition.
- Use the assessment in the Student Edition to gauge student understanding of the content.

Part 1: Show Them How

Review analog and digital clocks with students, including the meaning of the hands on an analog clock.

Explain to students that they will build a linear clock, using it much like a timeline to use skips or jumps to determine elapsed time.

Build a linear clock with students like the one illustrated. *Note*: For the sake of time and consistency, you may want to build this for the students before the lesson, rather than having students build it along with you. Directions for building the linear clock are in Chapter 3, "Introduction to Time." Use red 1x1 bricks as time markers for the problems in this chapter.

Begin by making sure students know how to read the linear clock. Touch the brick representing each hour on the clock face and say the hour that the brick represents. Begin at 12:00 am, and work your way around the baseplate, ending at 12:00 pm.

Make sure students understand that the distance between side-by-side studs on the baseplate represents a 5-minute interval. Practice counting by 5-minute intervals, 10-minute intervals, and half-hour intervals by touching the studs that represent each time.

Problem #1: Jodi's party begins at 7:30 pm and ends at 10:00 pm. How long is her party?

Step 1: Find the start time (7:30 pm) on the linear clock, and place a red 1x1 brick on that stud.

Step 2: Find the end time (10:00 pm) on the linear clock, and place a red 1x1 brick on that stud.

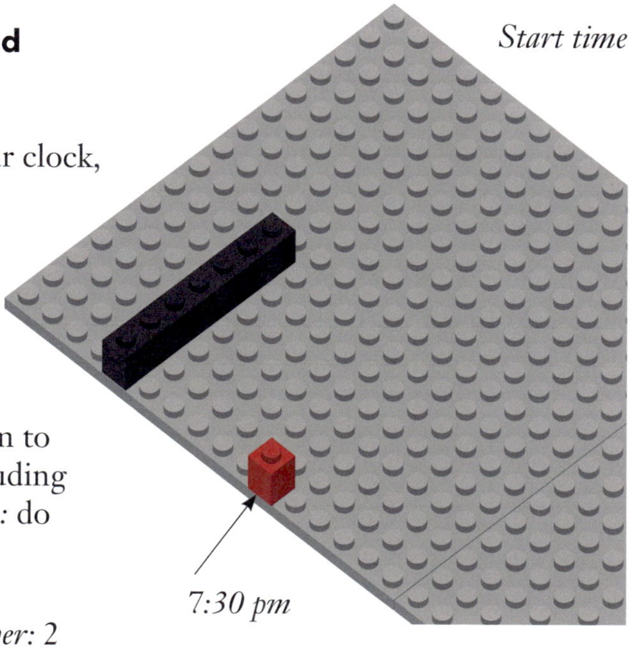

Step 3: Count in 5-minute intervals from 7:30 pm to 10:00 pm, by touching the studs in between (including the 10:00 stud) and counting aloud together. *Note:* do not count the stud representing the start time.

Step 4: Ask students the length of the party (*answer:* 2 hours and 30 minutes).

Step 5: Have students draw the part of the clock that shows the start time and end time, and label their drawings to show the solution.

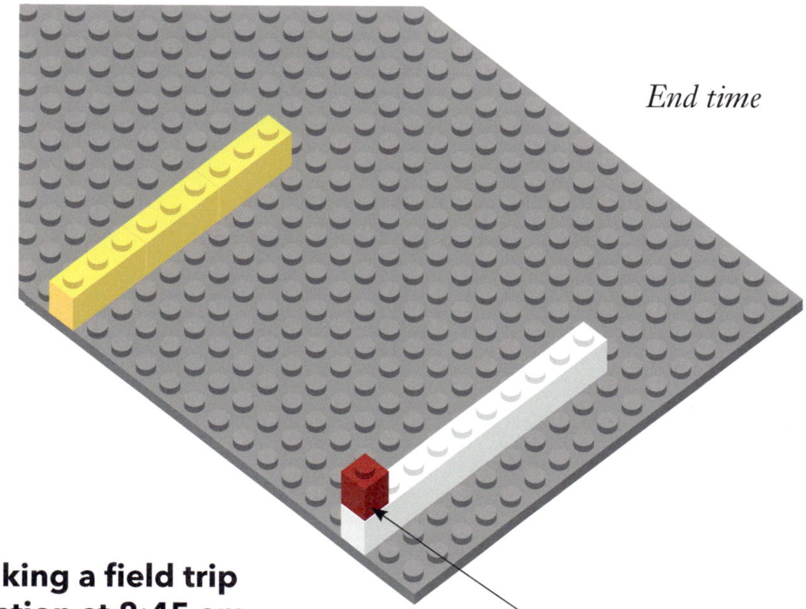

Problem #2: Imagine that you are taking a field trip on a train. The train will leave the station at 8:45 am and return to the station at 1:35 pm. How much time will the ride take?

Step 1: Find the start time (8:45 am) on the linear clock and place a red 1x1 brick on that stud.

Step 2: Find the end time (1:35 pm) on the linear clock and place a red 1x1 brick on that stud.

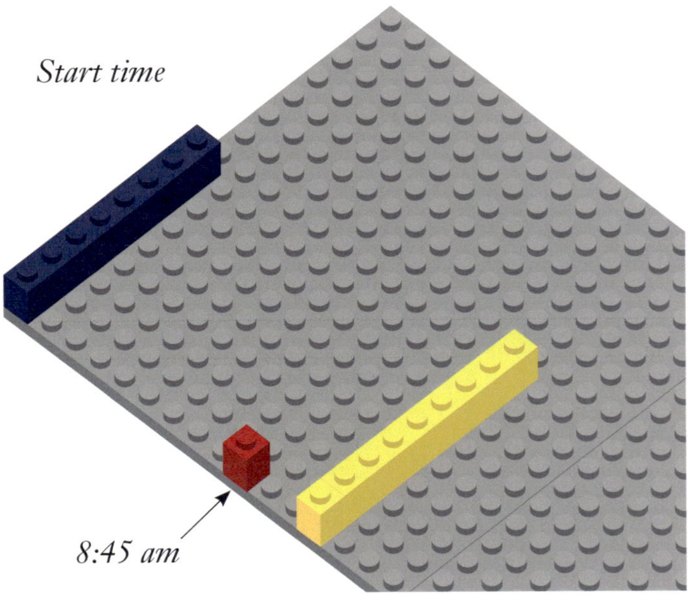

Step 3: Count up in 5-minute intervals from 8:45 to 1:35, by touching the studs in between (including the 1:35 stud) and counting aloud together.

Step 4: Ask students how much time the train ride will take (*answer:* 4 hours and 50 minutes).

Step 5: Have students draw the part of the clock that shows the start time and end time, and label their drawings to show the solution.

Note: Make sure students understand that they do not count the stud representing the start time, but they do count the stud representing the end time. This can be confusing for young students when learning to measure elapsed time.

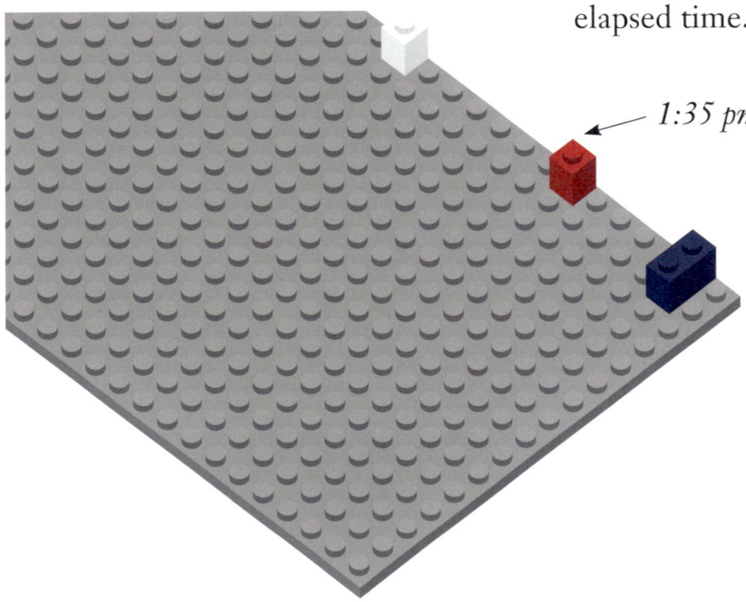

Problem #3: John gets up for school at 6:30 am. It takes him 10 minutes to brush his teeth, 15 minutes to get dressed, and 35 minutes to eat breakfast. His bus comes at 7:40 am. Will he make it in time?

Step 1: Find the start time (6:30 am) on the linear clock, and place a red 1x1 brick on that stud.

Step 2: Count 10 minutes past 6:30 am and place a second red 1x1 brick at the time (6:40 am).

Step 3: Count 15 past 6:40 am and place a third red 1x1 brick at the time (6:55 am).

Step 4: Count 35 past 6:55 am and place a fourth red 1x1 brick at the time (7:30 am).

Step 5: Have students draw the part of the clock that shows the start time and end time, and label their drawings to show the solution. Ask students to state what time John will be ready, and whether he will make his bus (*answer:* He will be ready at 7:30 am and the bus leaves at 7:40 am, so he will make it!)

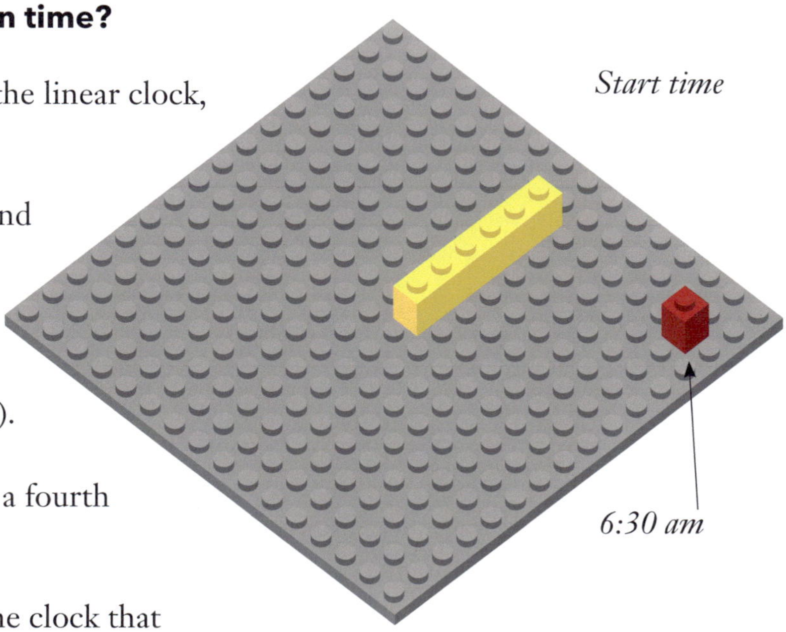

Start time

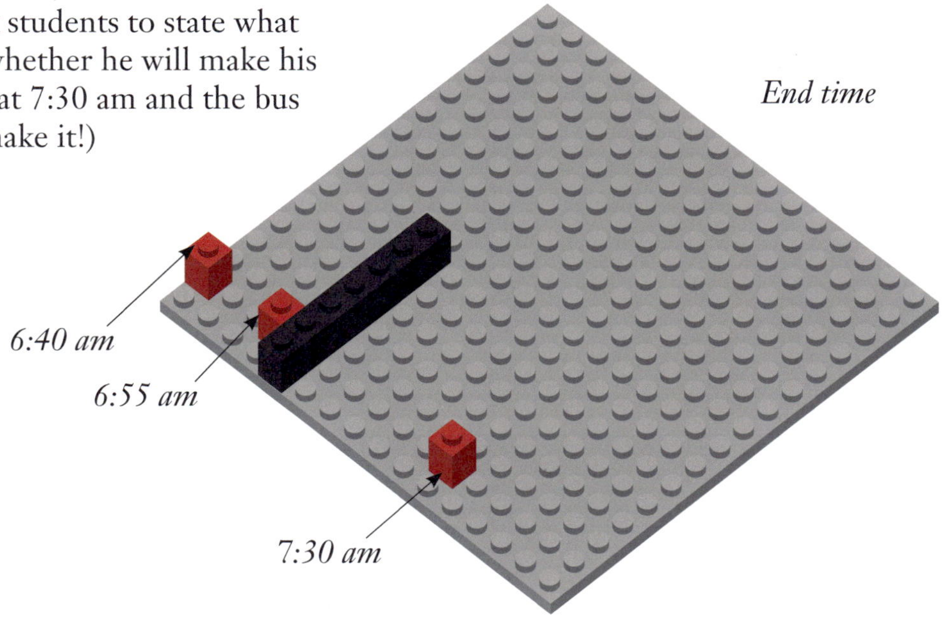

End time

Problem #4. The football team is leaving for their game at 3:15 pm. The game begins at 7:30 pm. If they arrive at 4:00 pm, how much time will they have to practice before the game starts?

Step 1: Place a red 1x1 brick at 3:15 pm to mark the start time.

Step 2: Place a second red 1x1 brick at 4:00 pm. *Note*: Students should recognize that 4:00 pm is the starting point for the problem, and that 3:15 pm is extraneous information. Students should count the time between 4:00 pm and 7:30 pm to determine the team's practice time.

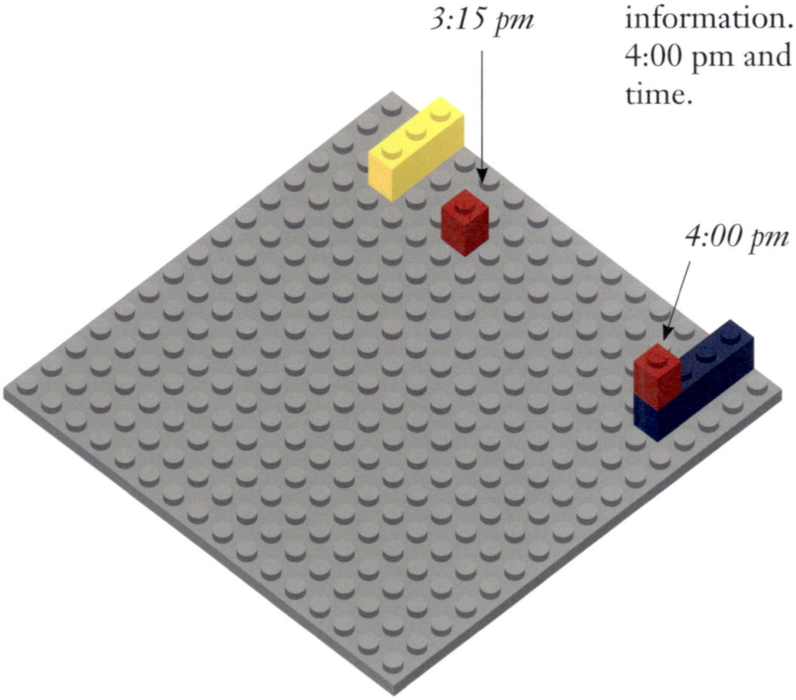

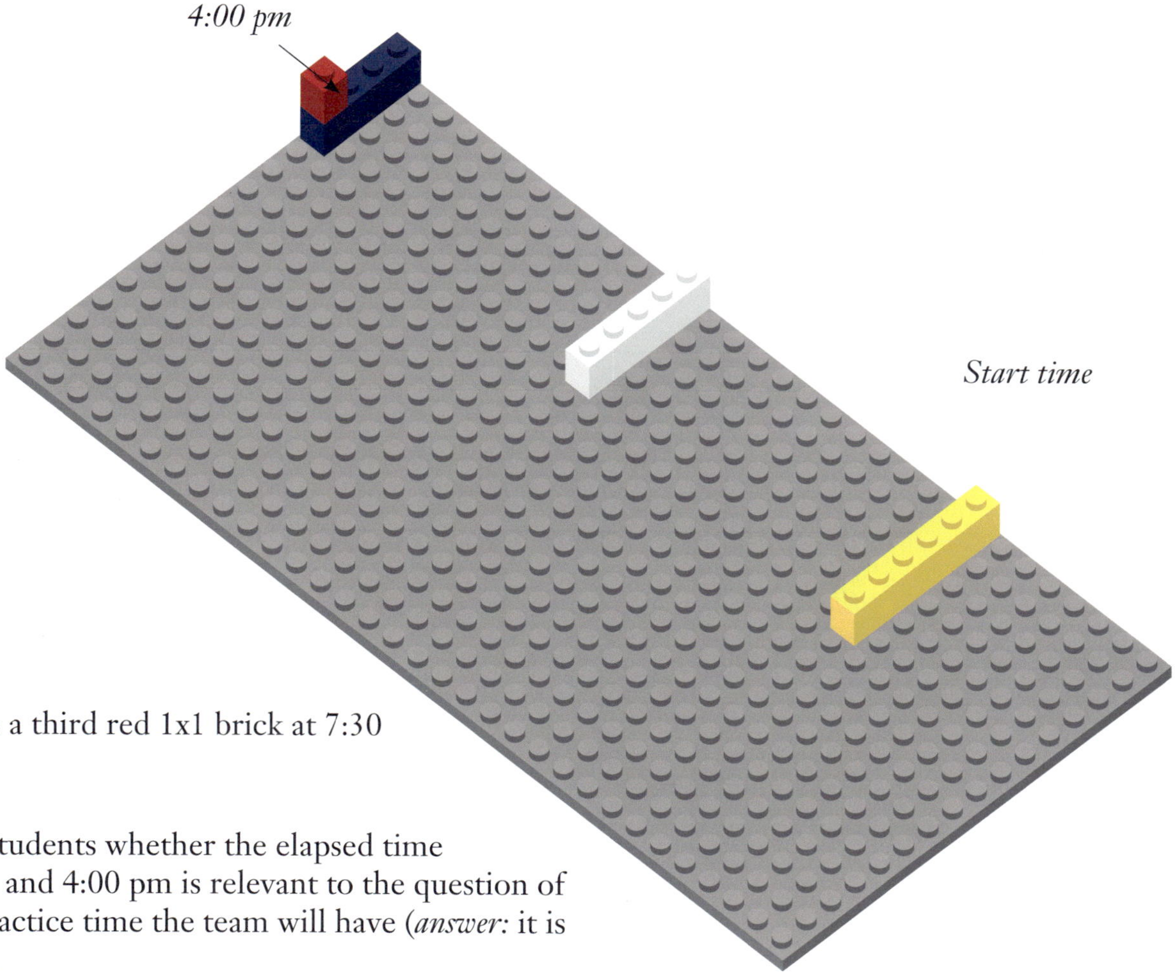

Start time

Step 3: Place a third red 1x1 brick at 7:30 pm.

Step 4: Ask students whether the elapsed time between 3:15 and 4:00 pm is relevant to the question of how much practice time the team will have (*answer:* it is not relevant).

Step 5: Count the time in 5-minute intervals from 4:00 pm until 7:30 pm (*answer:* 3 hours and 30 minutes). Have students draw the part of the clock that shows the start time and end time, and label their drawings to show the solution.

End time

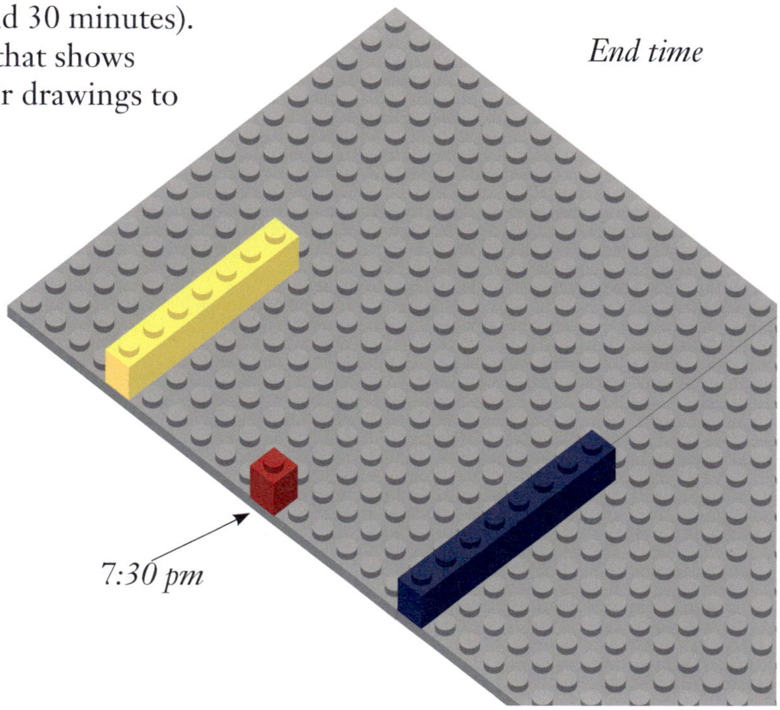

Part 2: Show What You Know

1. Playtime begins at 11:25 am. Lunch begins at 12:10 pm. How much time will pass between playtime and lunch? Draw the part of the clock that shows the start time and end time. Label your drawing to show the solution.

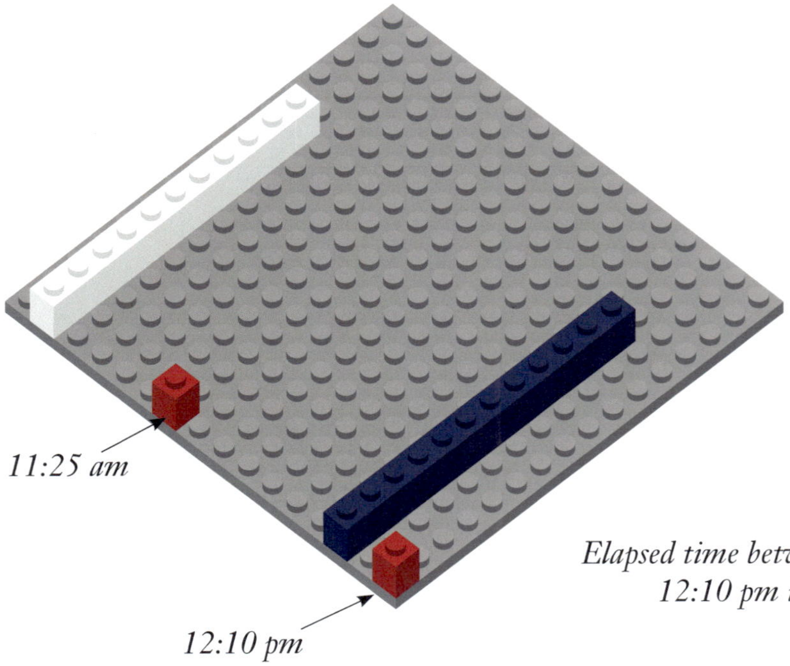

Elapsed time between 11:25 am and 12:10 pm is 45 minutes

2. School begins at 8:05 am. The school day is 6 hours and 30 minutes long. What time will it be when school ends? Draw the part of the clock that shows the start time and end time. Label your drawing to show the solution.

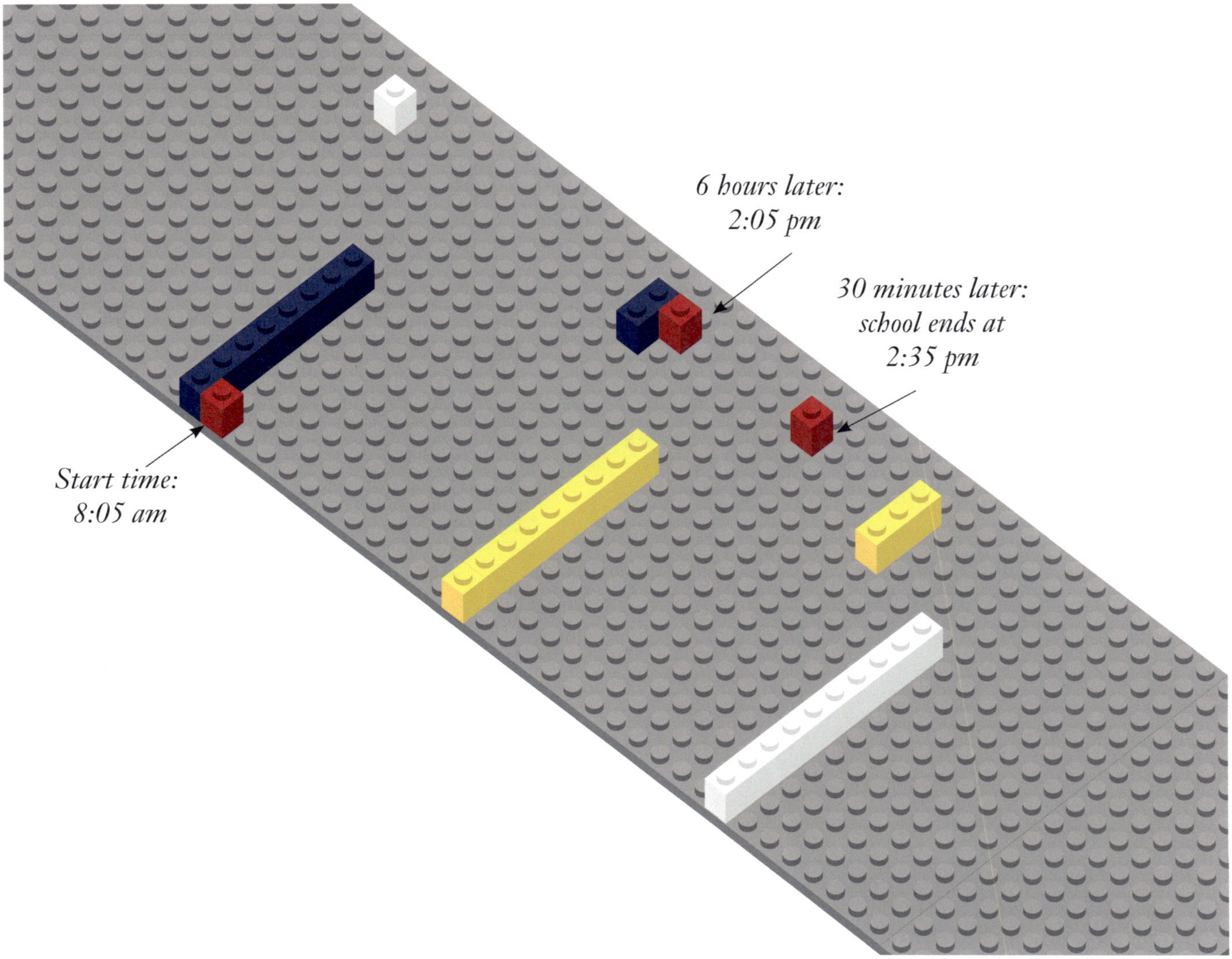

3. The recital begins at 2:30 pm and ends at 3:50 pm. How much time will the recital take? Draw the part of the clock that shows the start time and end time. Label your drawing to show the solution.

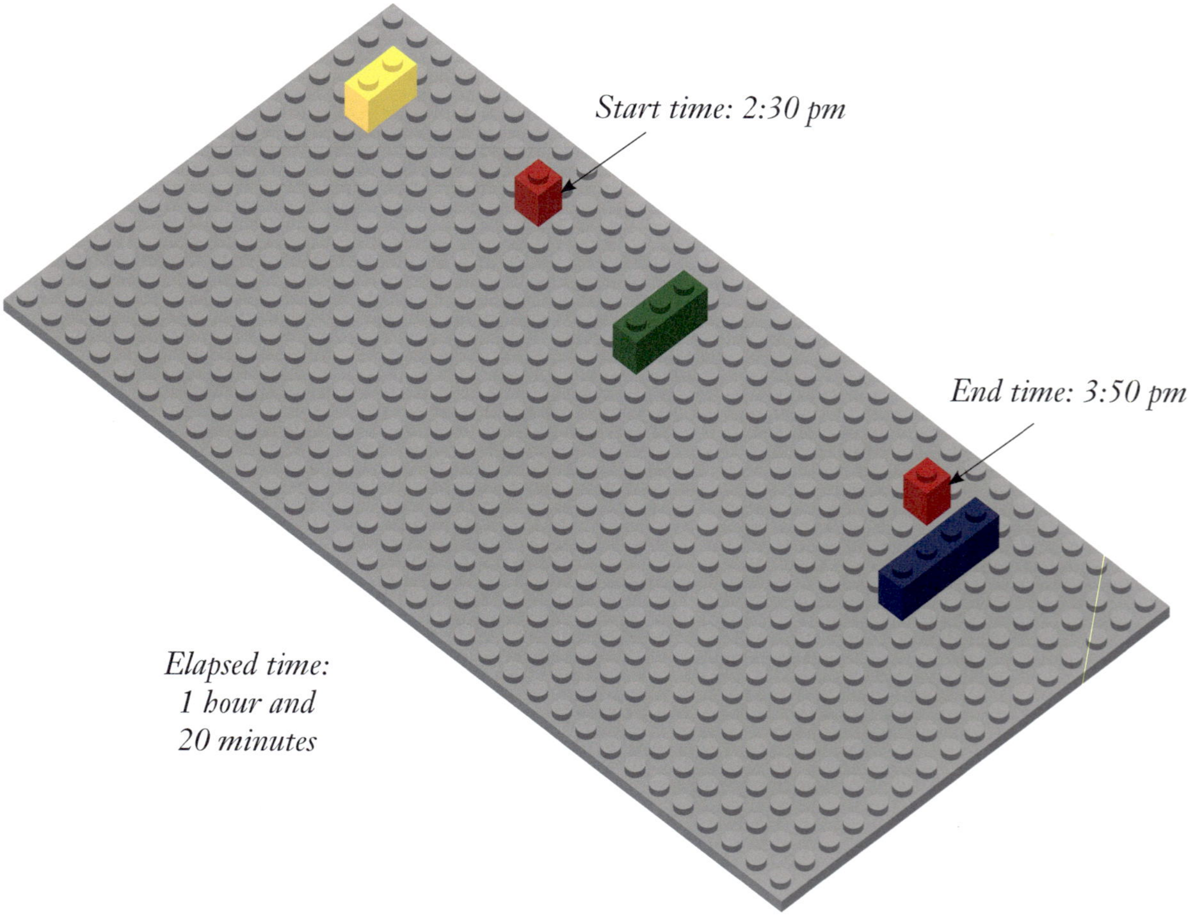

4. The morning session of the cheerleading competition is scheduled to start at 9:00 am. The lunch break is at 11:45 am. How long is the morning session? Draw the part of the clock that shows the start time and end time. Label your drawing to show the solution.

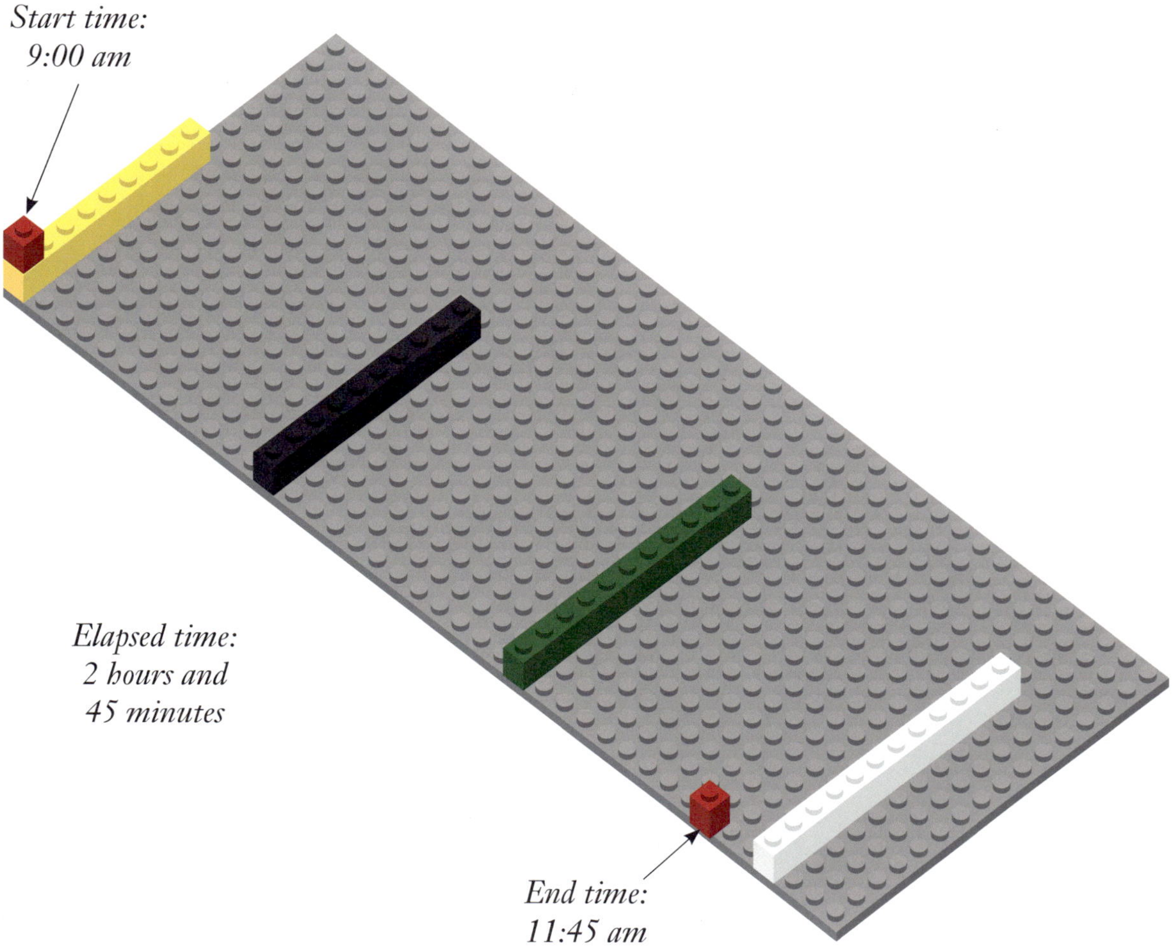

5

SUGGESTED BRICKS

Size	Number
1x1	24
1x2	16
1x4	8
1x8	2
1x16	2

Note: Using a baseplate helps keep the bricks in place. One baseplate is suggested for these activities.

CUSTOMARY LIQUID MEASURES

Students will learn/discover:
How to determine the number of quarts, pints, and cups in a gallon

Why is this important?
Measuring liquid amounts is a key skill in cooking and other activities of everyday life. The ability to convert between liquid measures helps students make connections to addition, subtraction, multiplication, and division. In higher grades students will learn about fractional parts of measures, and this is the foundation on which they will build that knowledge.

Vocabulary:
- Customary measure: The US measurement system
- Gallon: The largest unit of customary liquid measure
- Quart: $1/4$ of a gallon
- Pint: $1/2$ of a quart
- Cup: $1/2$ of a pint

How to use the companion student book, *Basic Measurement Using LEGO® Bricks–Student Edition*:
- After students build their models, have them draw the models and explain their thinking in the Student Edition. Recording the models on paper after building them with bricks helps reinforce the concepts being taught.
- Discuss the vocabulary for each lesson with students as they work through the Student Edition.
- Use the assessment in the Student Edition to gauge student understanding of the content.

Part 1: Show Them How

Show students the following bricks, explaining that each brick represents a unit of customary liquid measure in this lesson:

1x16 brick = 1 gallon

1x4 brick = 1 quart

1x2 brick = 1 pint

1x1 brick = 1 cup

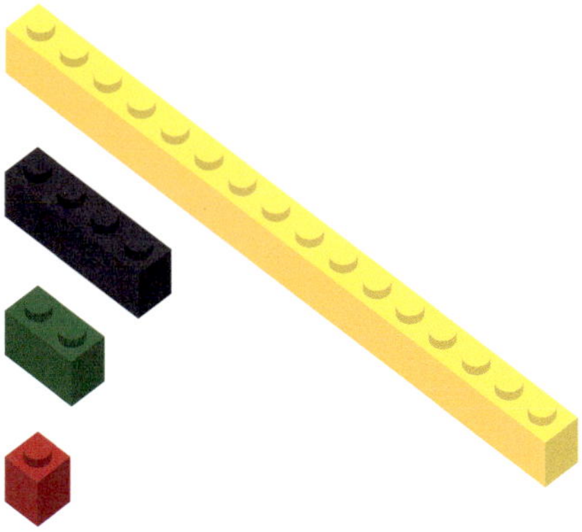

1. Show students a 1x16 brick. Explain that this brick represents 1 gallon. Place the 1x16 brick horizontally at the top of a baseplate. Have students create the same model along with you.

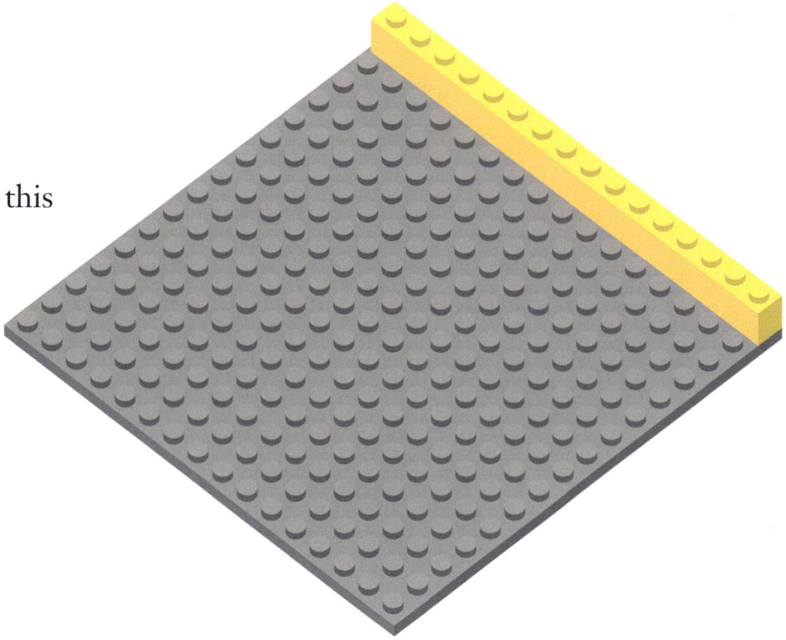

2. Explain to students that there are 4 quarts in 1 gallon. Ask students to find a brick to represent 1 quart (*answer:* 1x4 brick).

Add the bricks that represent 4 quarts to the model by placing a row of 1x4 bricks below the 1x16 brick. Have students add the same to their models.

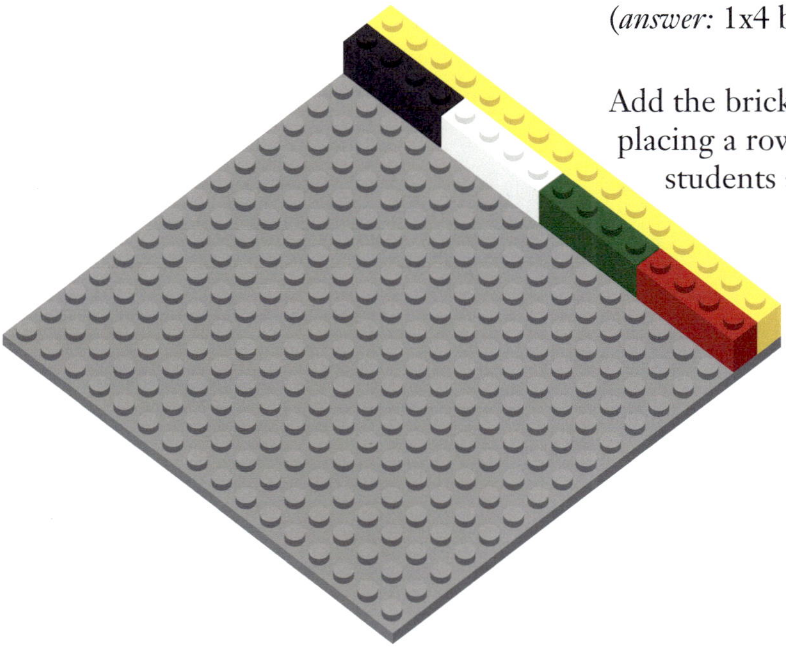

3. Explain to students that there are 2 pints in each quart. Ask students to find a brick to represent 1 pint (*answer:* 1x2 brick). Add the bricks that represent 2 pints for each quart by placing a row of 1x2 bricks below the row of 1x4 bricks. Have students add the same to their models.

Ask students how many pints are in a gallon (*answer:* 8 pints).

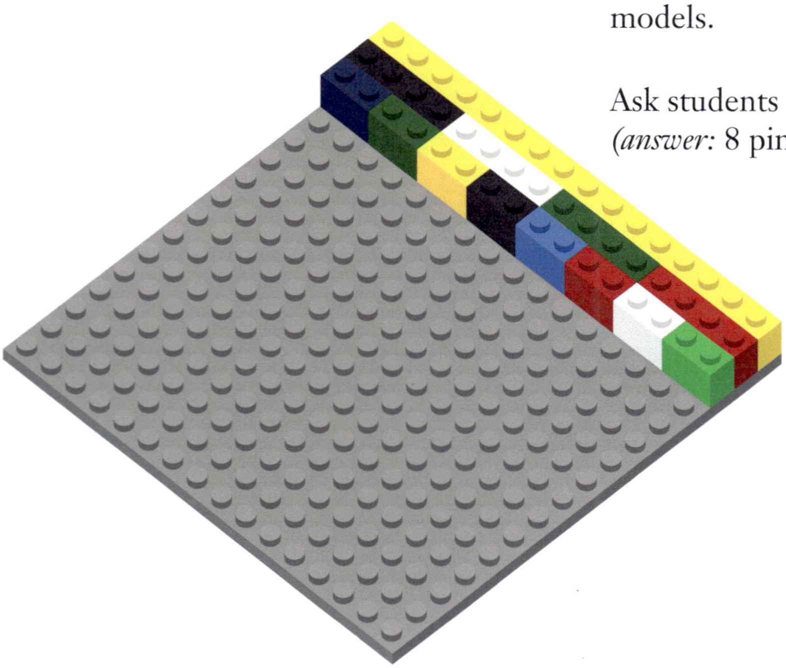

4. Explain to students that there are 2 cups in each pint. Ask students to find a brick to represent 1 cup (*answer:* 1x1 brick). Add cups to the gallon model by placing a row of 1x1 bricks below the row of 1x2 bricks. Have students add the same to their models.

Ask students how many cups are in 1 gallon (*answer:* 16 cups).

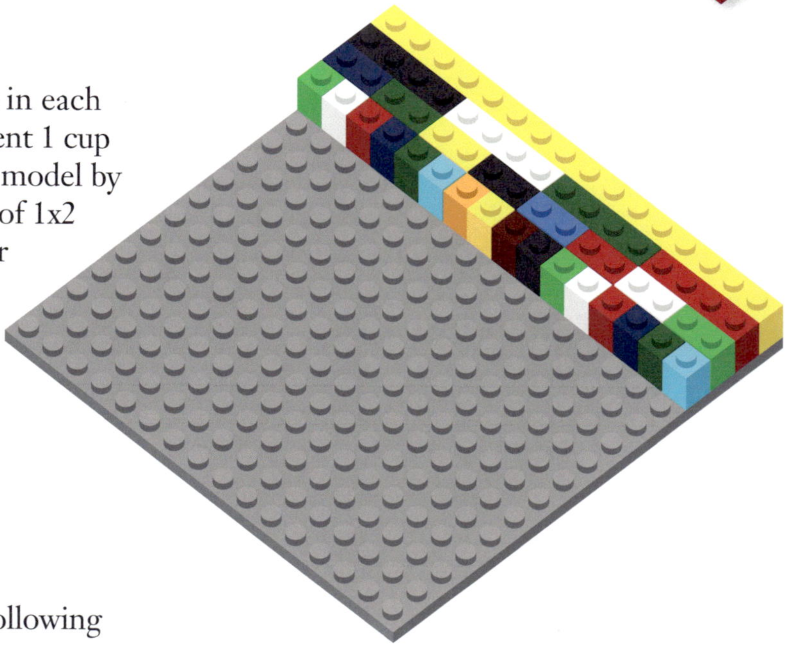

5. Have students use this model to solve the following problems:

a. Count the bricks to determine how many quarts are in 2 gallons (*answer:* 8 quarts). *Note:* Students can use one-to-one correspondence, counting the four 1x4 bricks twice to arrive at the answer.

b. Build a model to show the number of pints in 3 quarts (*answer:* 6 pints). *Note:* Students should use three 1x4 bricks to show 3 quarts, then show 2 pints in each quart with two 1x2 bricks below each 1x4 brick.

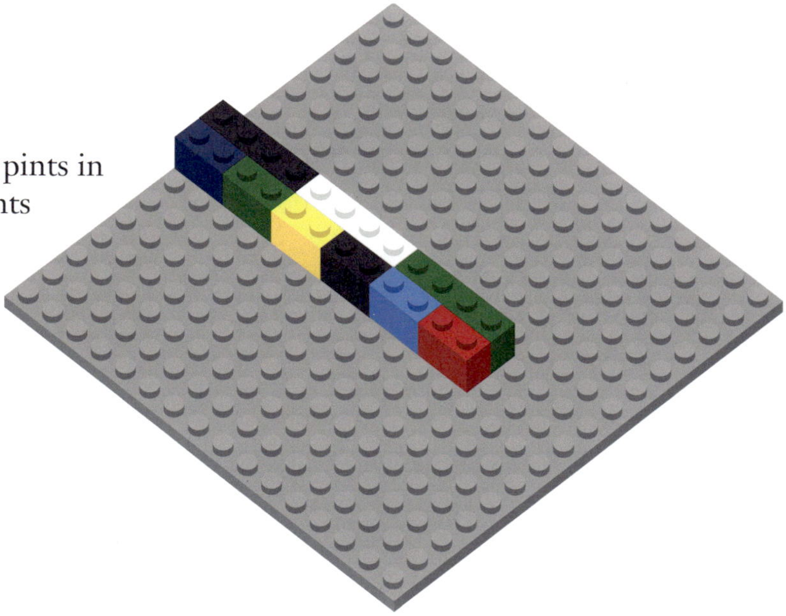

c. Build a model to show the number of pints in 8 cups (*answer:* 4 pints).

Note: Students should model 8 cups with eight 1x1 bricks, then model pints by placing four 1x2 bricks below them (each 1x2 brick should line up with two 1x1 bricks).

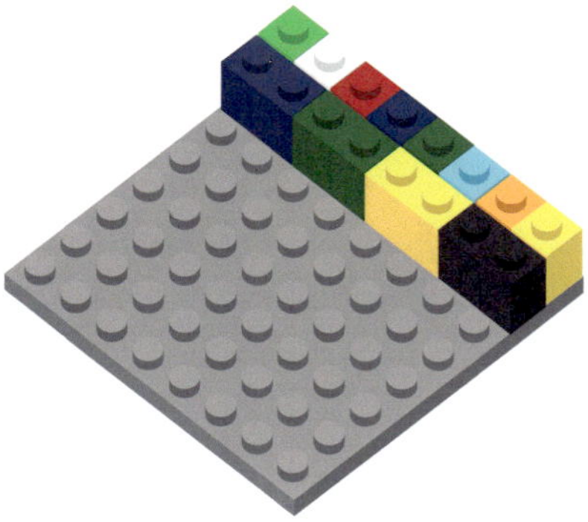

Part 2: Show What You Know

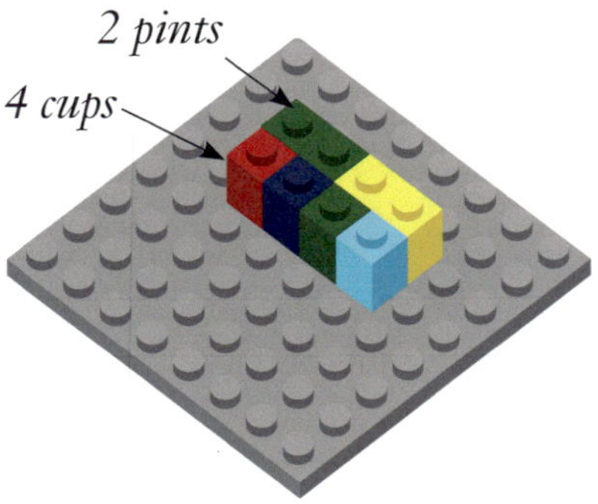

2 pints = 4 cups

1. Can you build a model to show how many cups are in 2 pints? Draw your model and label the bricks used to show cups and pints.

Note: Have students create a chart for their data collection as shown.

	Brick Size	Number
pints	1x2	2
cups	1x1	4

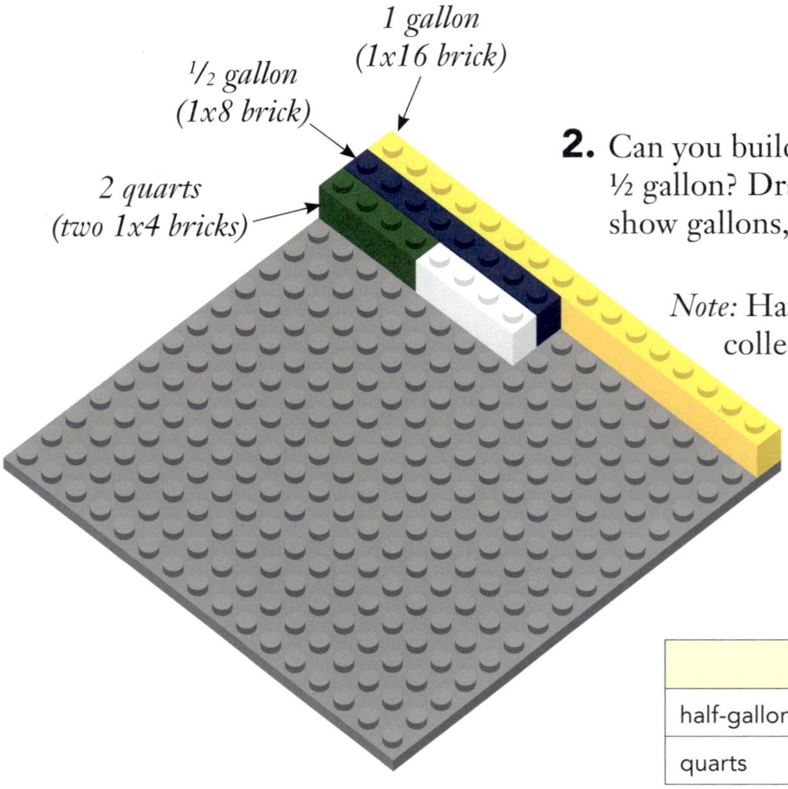

2. Can you build a model to show how many quarts are in ½ gallon? Draw your model and label the bricks used to show gallons, half-gallons, and quarts.

Note: Have students create a chart for their data collection as shown.

Hint: First, model ½ gallon using 1x8 bricks. How many 1x8 bricks are used to model ½ gallon?

	Brick Size	Number
half-gallons	1x8	1
quarts	1x4	2

3. Can you build a model to show how many pints are in 3½ quarts? Draw and explain your model.

Possible model: 3 quarts represented by three 1x4 bricks, with one 1x2 brick showing an additional ½ of 1 quart (or one pint). The row of 1x2 bricks shows 7 pints in 3½ quarts.

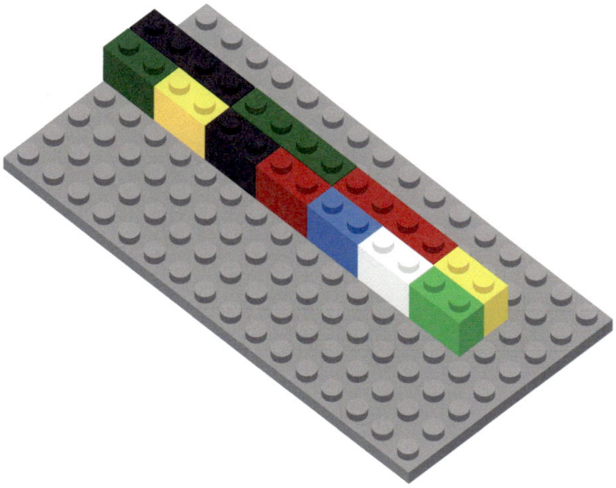

4. If you have 1 quart and 1 pint, how many cups do you have? Build a model to show your solution. Draw and explain your model.

Possible model: 1 quart represented by one 1x4 brick and 1 pint represented by one 1x2 brick. The row of 1x1 bricks shows 6 cups in 1 quart and 1 pint.

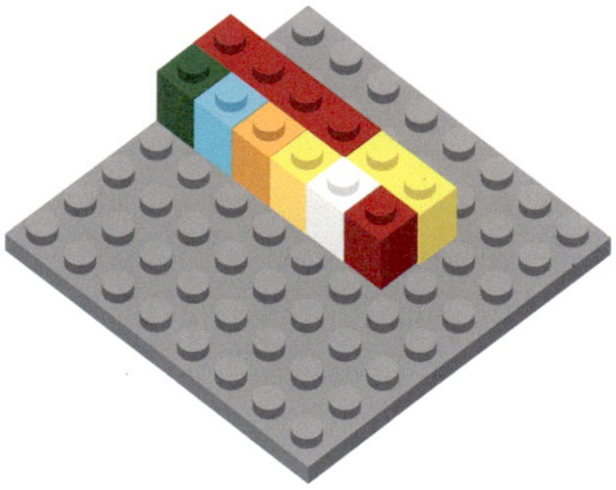

6

SUGGESTED BRICKS

Size	Number
1x1	24–30
1x2	14
1x3	10
1x4	8

Note: Using a baseplate helps keep the bricks in place. One baseplate is suggested for these activities.

METRIC MEASURES

Students will learn/discover:

- To identify the units in the metric system
- How the metric system is related to the base ten system
- How place value is related to reading decimal numbers
- How to make basic conversions using the metric system

Why is this important?

The metric system of measure is a standard system utilized in most countries throughout the world. It is particularly important in math and science fields. To compete globally, students must be able to use this system to measure attributes including distance, mass, and liquid amounts. It is an easy system to manage since it is based on the decimal system that we use in the United States. If students have a firm understanding of place value, they can easily master the metric system. Place value, decimal computation, and expanded forms are all concepts utilized when converting within the metric system. These concepts are learned in grades 2–5, and should be reviewed prior to teaching the metric system to ensure students have a firm understanding of them.

Vocabulary:

- Base units of measure: Liter (liquid), meter (distance), and gram (mass)
- Convert: To change one unit of measure into another unit (e.g., meters to centimeters)
- Metric system of measurement: Standard system of measurement based on the decimal system
- Prefixes used in metric system:
    - Kilo (1000)
    - Hecto (100)

- Deka (10)
- Deci (1/10)
- Centi (1/100)
- Milli (1/1000)

How to use the companion student book, *Basic Measurement Using LEGO® Bricks–Student Edition*:

- After students build their models, have them draw the models and explain their thinking in the Student Edition. Recording the models on paper after building them with bricks helps reinforce the concepts being taught.
- Discuss the vocabulary for each lesson with students as they work through the Student Edition.
- Use the assessment in the Student Edition to gauge student understanding of the content.

Part 1: Show Them How

Note: This chapter will not require much drawing of models due to their complexity. Students should collaborate and share using the math vocabulary.

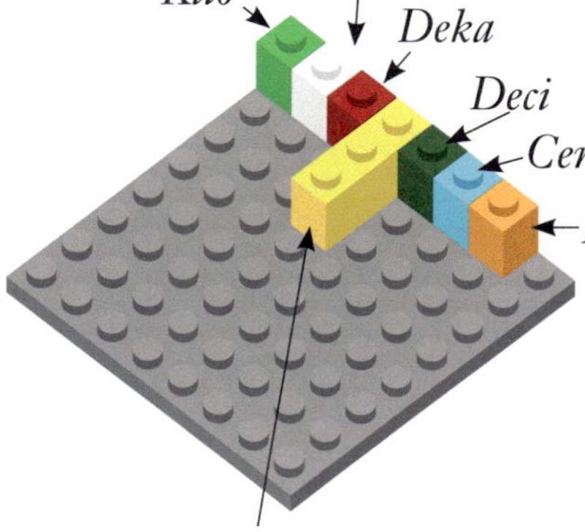

Have students build a model of the metric system as illustrated, then draw and label the model.

Explain that each brick in the model represents a unit of measure in the metric system. Describe the amount that each brick represents as a multiple of the base unit (left to right): kilo (x 1000), hecto (x 100), deka (x 10), base (liter, meter, gram), deci (x 1/10), centi (x 1/100), milli (x 1/1000). *Note:* the base unit is represented by a 1x3 brick, and all other units are represented by 1x1 bricks.

Once students are familiar with the metric system model, review how to use bricks to model place value:

1x1 = 0-9
1x2 = 10s
1x3 = 100s
1x4 = 1000s

Note: See *Addition Using LEGO Bricks—Teacher Edition* to review place value models.

Using Multiplication to Convert Metric Measures
Explain to students that in metric measure conversion, each move to the right on the model signifies multiplication by 10 and each move to the left signifies division by 10.

Problem #1: Ask students to identify the number of centimeters in 1 meter (*answer:* 100 centimeters).

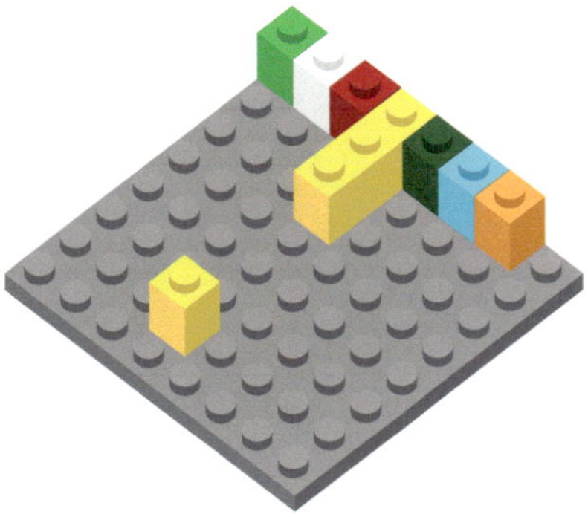

Step 1: Place a 1x1 brick under the base unit (meter) brick to show the starting place of 1 meter.

Step 2: Ask students how far away in bricks the centimeter value is from the base meter value (*answer:* 2 bricks to the right). *Note:* because students are moving to the right, each brick represents multiplication by 10.

Step 3: Ask students to find a place value brick that represents 1 meter multiplied by 10 (*answer:* 1x2 brick) Place this brick under the decimeter brick in the model. This new brick represents 10 decimeters.

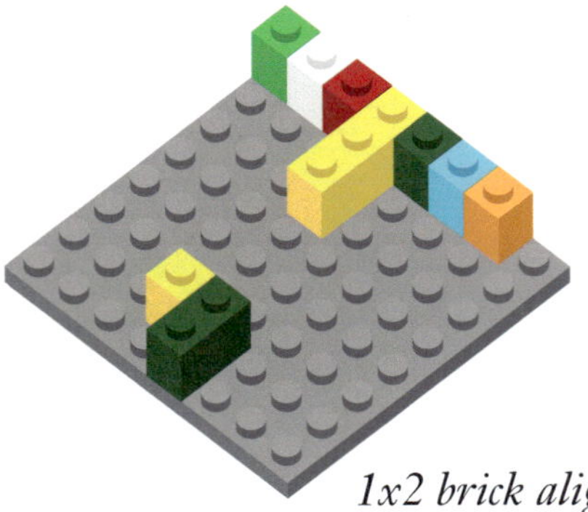

1x2 brick aligned under the decimeter brick shows that 1 meter is equivalent to 10 decimeters

Step 4: Ask students to find a brick that represents 10 decimeters multiplied by 10 (*answer:* 1x3 brick). Place this brick, which represents 100 centimeters, aligned under the centimeter brick in the model. Have students draw and label this model.

1x3 brick aligned under the centimeter brick shows that 1 meter is equivalent to 100 centimeters

Problem #2: Ask students to identify the number of grams in 21 kilograms (*answer:* 21,000 grams).

Step 1: Find the starting place on the metric model (*answer:* the far left brick, which represents kilos). Ask students which bricks can be used to represent 21 kilograms (*answer:* two 1x2 bricks and one 1x1 brick). Stack two 1x2 bricks and one 1x1 brick, aligning them under the kilo brick on the metric model.

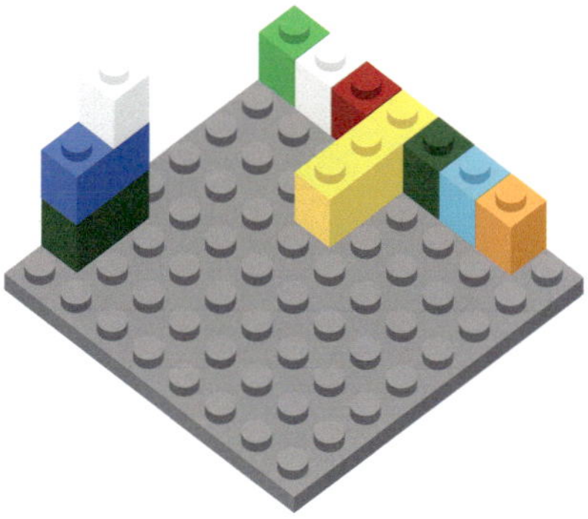

Step 2: Ask students to count the number of times they need to multiply by 10 to convert kilograms to grams (*answer:* three times). *Note:* to arrive at the answer, students should count the number of times they need to move to the right on the model to arrive at the brick that represents grams.

Step 3: Model the multiplication of 21 kilograms by 10 to show hectograms by placing a second stack of bricks underneath the hecto brick in the metric model. Ask students which bricks can be used to show the number of hectograms in 21 kilograms (*answer:* 21 kilograms is equal to 210 hectograms; this can be shown with two 1x3 bricks, representing 100 hectograms each, and one 1x2 brick, representing 10 hectograms).

21 kilograms = 210 hectograms

Step 4: Model the multiplication of 210 hectograms by 10 to make dekagrams, by placing a third stack of bricks underneath the deka brick in the metric model. Ask students which bricks should be used to show the number of dekagrams in 210 hectograms (*answer:* 2,100 dekagrams; shown with two 1x4 bricks, representing 1000 dekagrams each, and one 1x3 brick, representing 100 dekagrams).

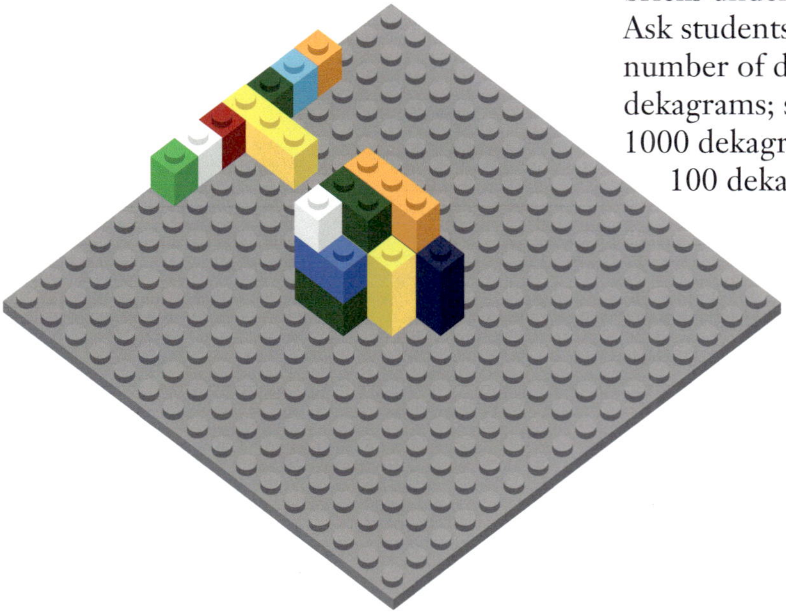

210 hectograms =
2100 dekagrams

Step 5: Model the multiplication of 2100 dekagrams by 10 to find the total number of grams in 21 kilograms. Place a fourth stack of bricks underneath the base unit (gram) brick in the model. Ask students which bricks should be used to show the number of grams in 2100 dekagrams (*answer:* 21,000 grams; shown with two sets of bricks in 1x5 groups, representing 10,000 grams each, and one 1x4 brick, representing 1,000 grams).

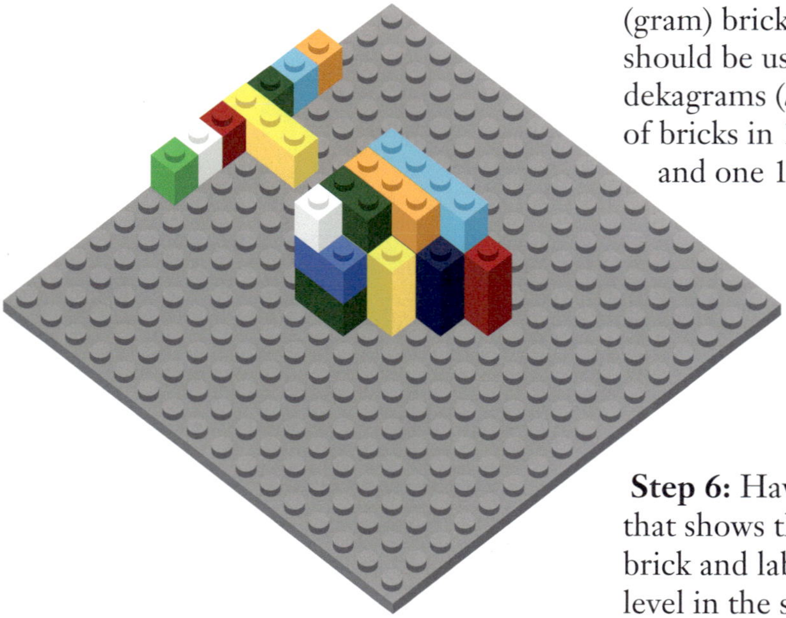

2100 dekagrams = 21,000 grams
21 kilograms = 21,000 grams

Step 6: Have students notice the pattern in the model that shows the multiplication. Ask students to draw each brick and label the number of bricks included at each level in the solution. Have students explain their thinking to a partner.

Using Division to Convert Metric Measures

Note: To model division with metric measures, the bricks are used to model decimal place values, rather than whole number place values. Demonstrate decimal place value modeling before starting to teach using division to convert metric measures.

Decimal Place Value Modeling:

Step 1: Explain to students that the bricks will now be used to represent place values of decimals, not whole numbers. *Note:* It is important to point out that students will be using place value modeling to show the values, as they did with whole numbers. The decimal place value model is essentially the reverse of the whole number model. *Note:* the 1x1 brick is not used, except when representing mixed decimals. In these models, the 1x1 brick serves as the decimal point.

Step 2: Build the decimal place value model for tenths, hundredths, and thousandths. Show students the model and have students create their own models, then draw and label them.

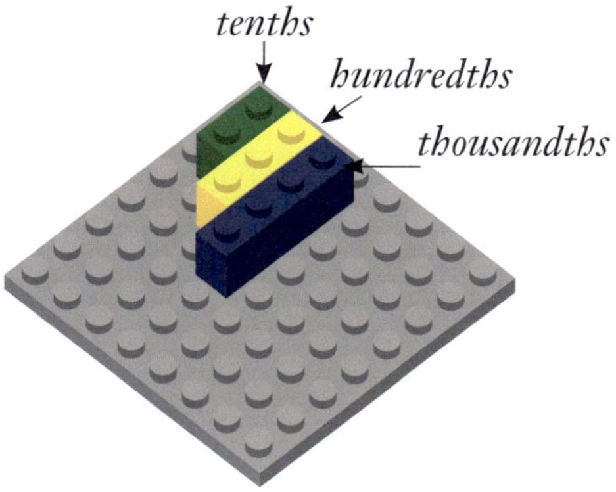

1x2 brick = tenths place
1x3 brick = hundredths place
1x4 brick = thousandths place

Step 3: Have students build and draw models for:

a) 0.32
b) 0.20
c) 0.324

a) *solution:*

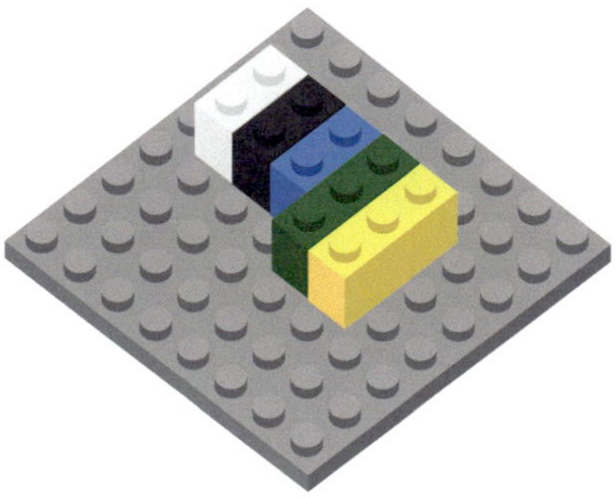

32 hundredths; or
(3 x $^1/_{10}$) + (2 x $^1/_{100}$); or
0.30 + 0.020; or
3 tenths (each 1x2 brick is 1 tenth) + 2 hundredths (each 1x3 brick is 1 hundredth)

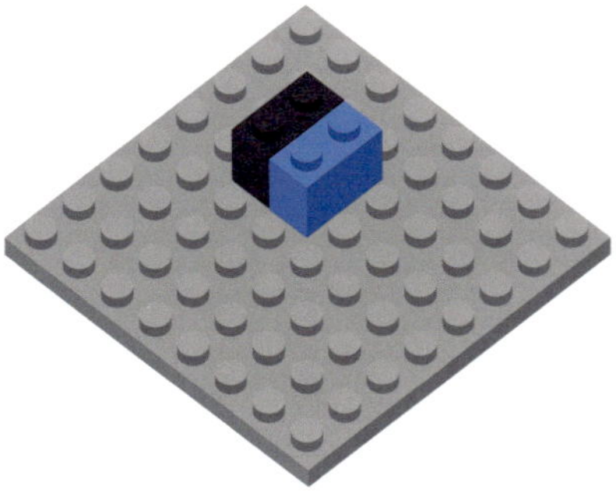

b) *solution:*

2 tenths; or
2 x $^1/_{10}$; or
0.20

c) *solution:*

324 thousandths; or
(3 x $^1/_{10}$) + (2 x $^1/_{100}$) + (4 x $^1/_{1000}$); or
0.300 + 0.020 + 0.004; or
3 tenths + 2 hundredths + 4 thousandths

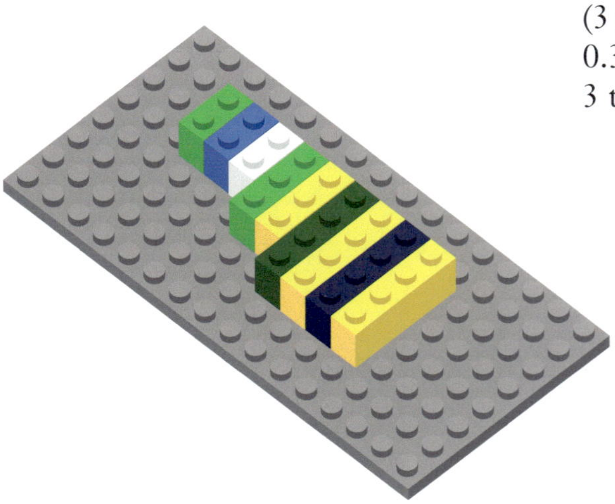

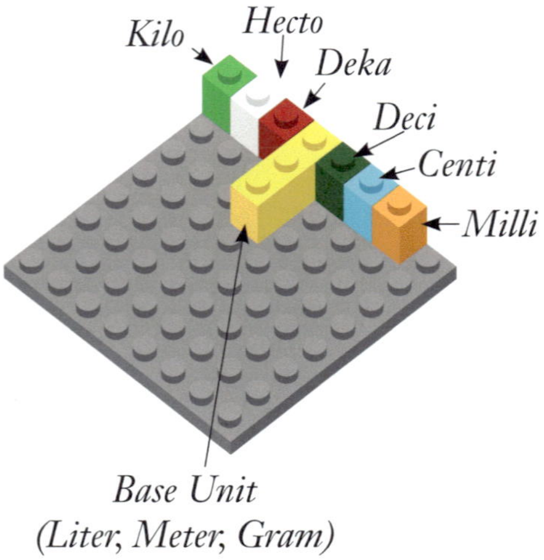

When you are sure that students understand how to model decimal place values, you can begin to model division with metric measures.

Review the model of the metric system. Students can refer to a drawing or an actual brick model.

Problem #1: How many meters are equal to 42 centimeters?

Note: Point out that this problem requires students to divide by 10, which is represented in the model by moving to the left. You will be using the method of modeling decimal place values.

Step 1: Place a 1x1 brick underneath the centimeter brick on the model, to show the starting point.

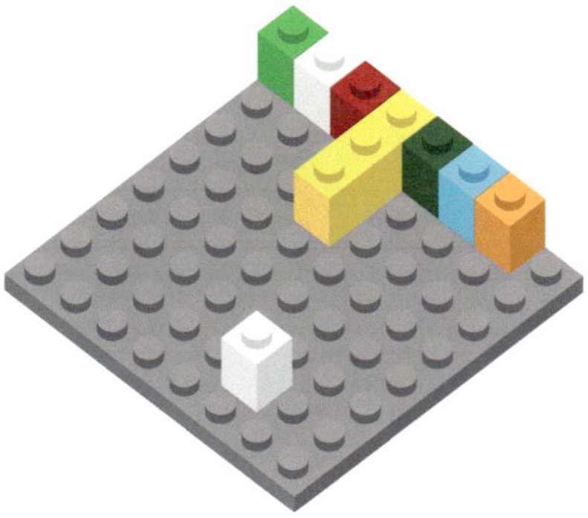

Step 2: Build a stack of bricks to represent 42 centimeters, using four 1x2 bricks (representing 4 tens or 4 sets of 10 centimeters), and two 1x1 bricks (representing 2 ones or 2 centimeters). Place the stack under the starting brick on the model.

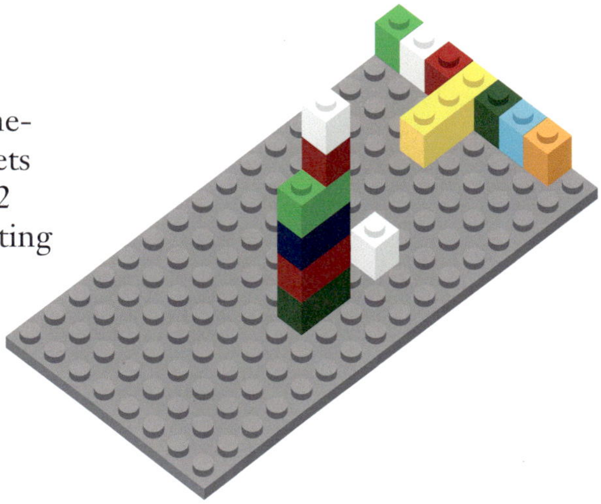

Step 3: Model the division of 42 centimeters by 10 to make decimeters. Separate the tens (four 1x2 bricks) from the ones (two 1x1 bricks) on the baseplate, and place a 1x1 brick between the stacks to represent a decimal point. This makes a visual representation of 42 divided by 10, or 4.2.

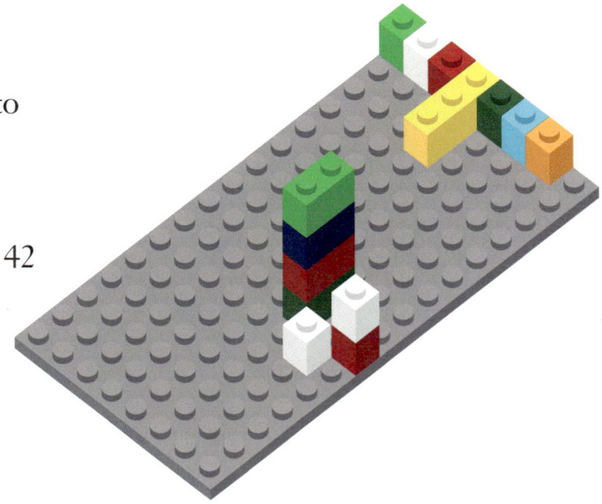

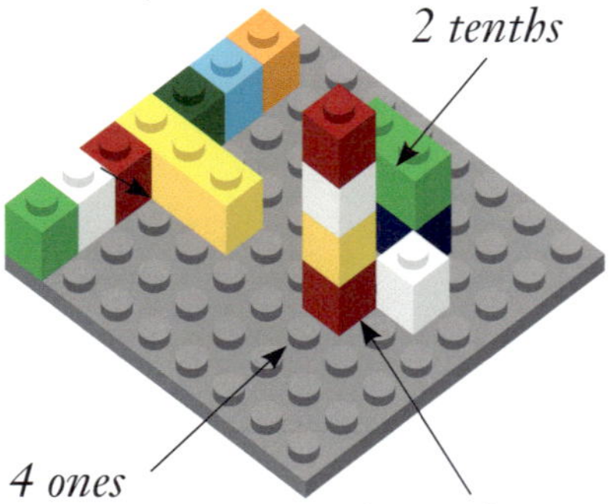

Now shift to using the bricks to represent decimal places. To model 4.2, align the "decimal point" brick under the decimeter place. Then build a stack of four 1x1 bricks (representing 4) to the left of the "decimal point," and build a stack of two 1x2 bricks (representing 2 tenths, or .2) to the right of the "decimal point." This models 42 centimeters divided by 10, or 4.2 decimeters.

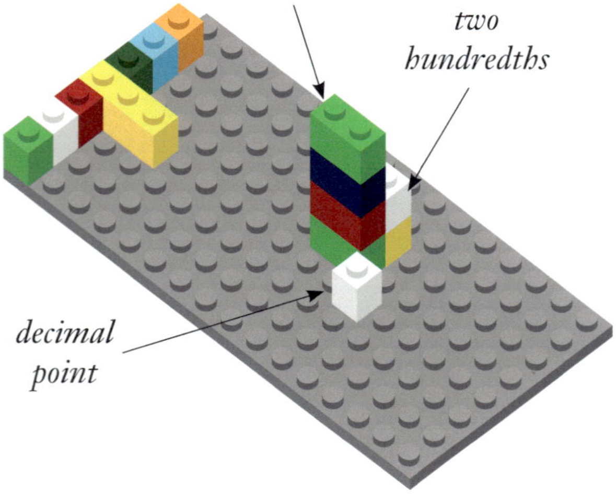

42 centimeters = 4.2 decimeters = .42 meters

Step 4: Model the division of 4.2 decimeters by 10 to find the equivalent in meters. Move the "decimal point" one stud, aligning it under the base unit meter brick. To show division by ten, replace the stack of four 1x1 bricks with a stack of four 1x2 bricks, representing 4 tenths, or .4, and move it to the right of the "decimal point." Replace the stack of two 1x2 bricks with a stack of two 1x3 bricks, representing 2 hundredths, or .02, and place them to the right of the tenths bricks. These bricks show .42 meters.

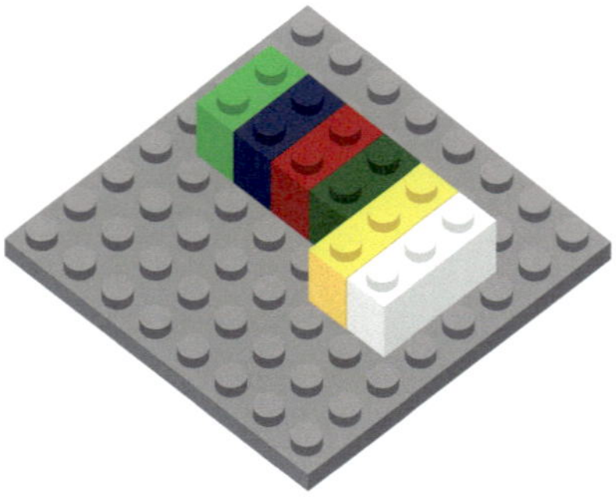

42 hundredths; or 4 tenths and 2 hundredths

Step 5: Unstack the model and display the place values in the number. Ask students to explain their thinking and justify their models with a partner. Have students display the final brick model in an unstacked format as shown, then draw and label their solution.

Show What You Know:

1. Can you build a model that shows 0.245? Draw and label the parts of your model, and write the expanded form of the number.

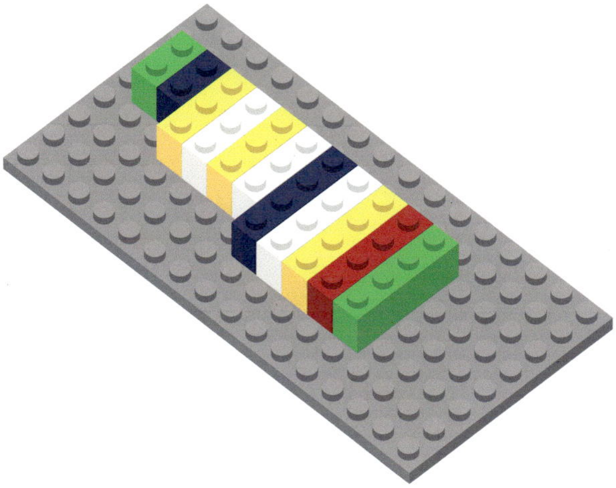

Possible solution:

2 hundred forty-five thousandths; or
$(2 \times {}^{1}/_{10}) + (4 \times {}^{1}/_{100}) + (5 \times {}^{1}/_{1000})$; or
0.200 + 0.040 + 0.005

2. Can you build a model to show the conversion of 3 meters to decimeters? Draw and label your model, and explain your thinking. *Hint:* you will be multiplying with whole numbers (moving to the right) in this problem.

Possible solution:

3 meters = 30 decimeters (3 x 10 decimeters)

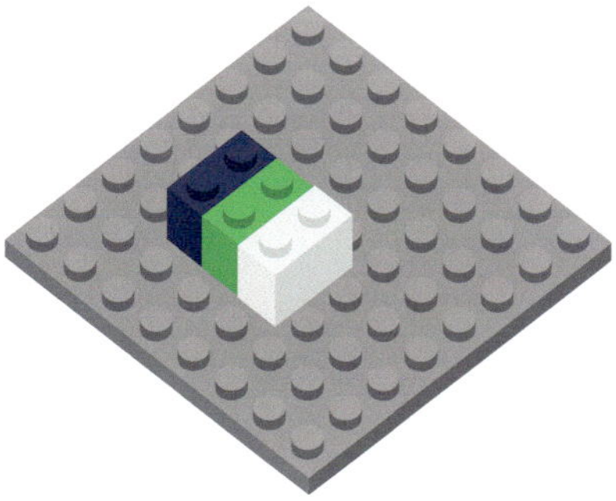

unstacked model of solution showing three 1x2 bricks representing 30 decimeters

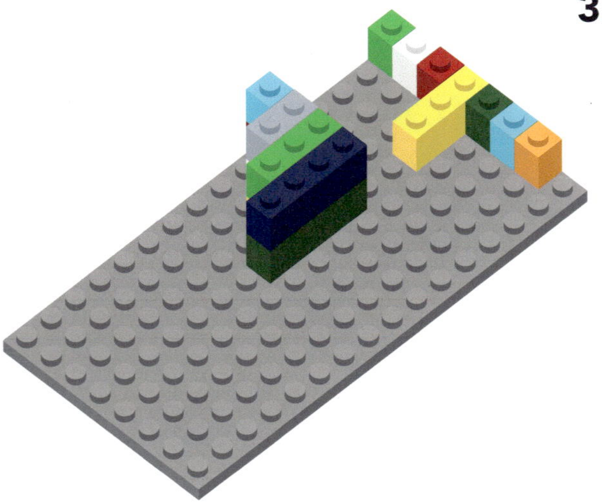

3. Can you build a model that shows the conversion of 2 kilograms to grams?

Draw and label your model, and explain your thinking.

Possible solution:

2 kilograms = 2000 grams (each move to the right represents multiplication by 10)

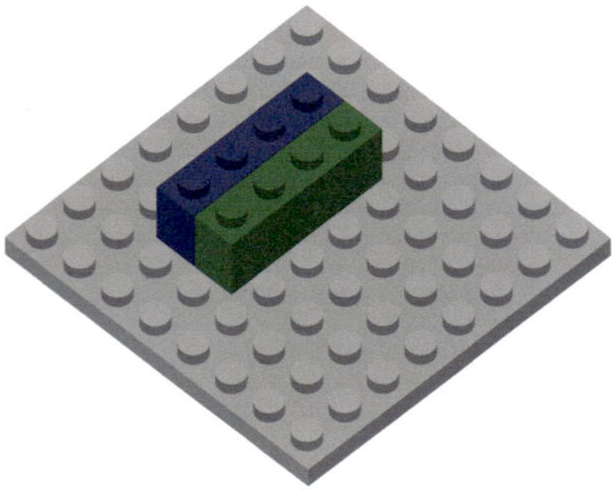

unstacked model of solution showing two 1x4 bricks representing 2000 grams

4. Can you build a model that shows the conversion of 4 liters to kiloliters?

Hint: This is a division problem. Because you are dividing each place by ten, the bricks will model decimal place values (e.g., the 1x4 brick represents thousandths).

Possible solution:

4 liters = 0.004 kiloliters (four thousandths of a liter)
4 L = 0.004 KL

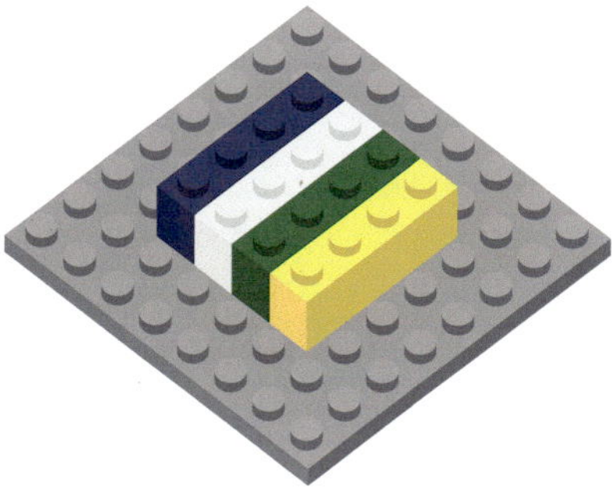

unstacked model of solution showing four 1x4 bricks representing four thousandths of a liter

7

SUGGESTED BRICKS

Size	Number
1x1	20
1x2	15
1x3	6
1x4	6
1x6	4
1x8	4
1x10	4
1x12	2
1x16	2

Note: Using a baseplate helps keep the bricks in place. Two baseplates are suggested for these activities.

PERIMETER

Students will learn/discover:

- How to determine the perimeter of rectangular and square areas
- How to use the mathematical formula for perimeter

Why is this important?

A basic understanding of perimeter is needed in everyday life. For instance, measuring perimeter is important when framing a picture or fencing around a garden. Students should use squares and rectangles to learn that perimeter is the distance around the outside of a shape. This topic should be explored as much as possible through real-world scenarios, not through the memorization of formulas.

Vocabulary:

- Area: The amount of space or the surface included within a set of lines
- Perimeter: The distance around the outside of a given shape; the sum of the measures of all the sides of a given shape
- Rectangle: A quadrilateral with four right angles and opposite sides equal (i.e., two longer parallel sides and two shorter parallel sides)
- Square: A quadrilateral with four right angles and all sides equal
- Formula: A standard way of showing a math procedure or sentence using numbers, that works in every situation

How to use the companion student book, *Basic Measurement Using LEGO® Bricks–Student Edition*:

- After students build their models, have them draw the models and explain their thinking in the Student Edition. Recording the models on paper after building them with bricks helps reinforce the concepts being taught.
- Discuss the vocabulary for each lesson with students as they work through the Student Edition.
- Use the assessment in the Student Edition to gauge student understanding of the content.

Part 1: Show Them How

Ask students to build a square using four 1x4 bricks and four 1x1 bricks, as well as a rectangle using two 1x10 bricks and two 1x3 bricks.

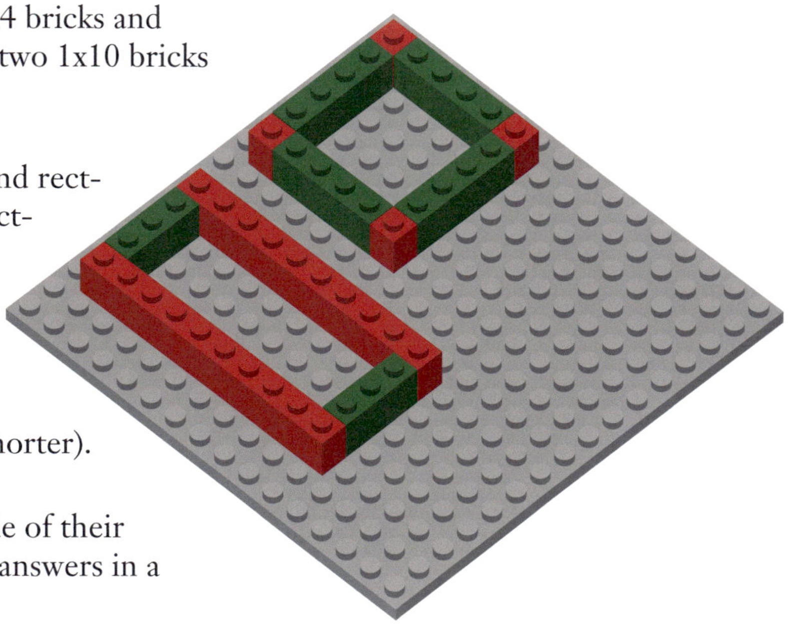

Discuss what students know about square and rectangular shapes. Explain that a square is a rectangle by definition, but a rectangle cannot be a square by definition. A square has four right angles and four sides of equal length; a rectangle has four right angles and opposite sides that are of equal length (but the sides may be two longer and two shorter).

Ask students to count the length of each side of their square in studs. Have students record their answers in a table.

Ask students to find the sum of all four side lengths by adding together the number of studs on each side. Explain that this sum is the *perimeter* of the square, or the distance in studs around the square.

Note: If students count the studs of the model built, they will need to **double count** the corner studs to find the perimeter of the rectangles. This is why they are asked to count the studs on each side and then add up all the sides to find the perimeter, rather than simply count the studs.

Sample Table for Square:

Side	Length in studs
1	5
2	5
3	5
4	5
Sum of all side lengths (perimeter)	20

Ask students to find the side lengths and perimeter of the rectangle in studs, and record their answers in another table. Explain that the addition of all the side lengths creates a formula that works every time for every rectangular shape.

Sample Table for Rectangle:

Side	Length in studs
1	10
2	5
3	10
4	5
Sum of all side lengths (perimeter)	30

Pose the following three problems for students to solve. To foster collaboration, have students work in groups.

Problem #1: Josh needs an outdoor pen for his new puppy. The pen must fit in his backyard, and it needs to be a rectangle. The perimeter of the pen can only be 42 feet.

Have students model a pen for the dog with bricks (1 stud = 1 foot), then draw the model and prove that the perimeter is 42 feet. Have students record the side lengths of the pen in a table and write a math sentence for the solution.

Have students or groups experiment with creating the pen in as many different ways as possible.

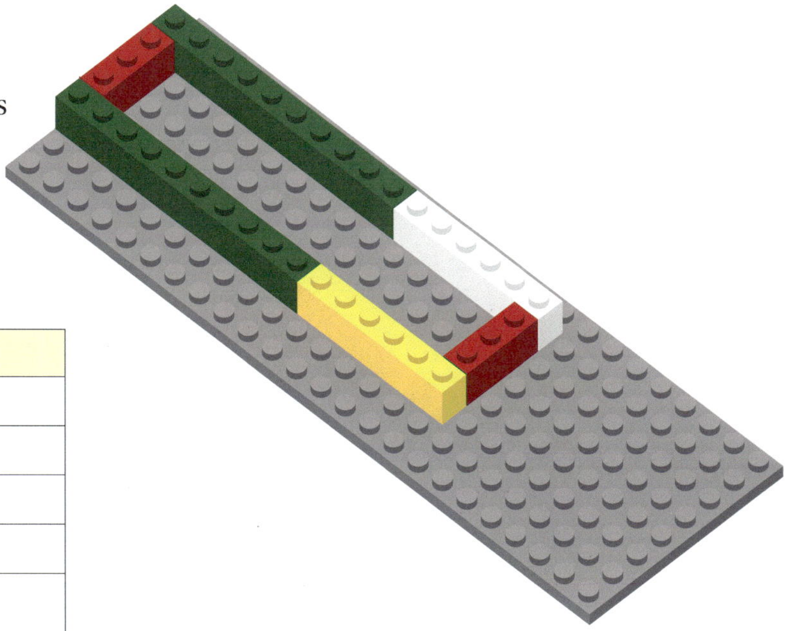

Possible solution:

Side	Length in studs
1	16 studs
2	5 studs
3	16 studs
4	5 studs
Sum of all side lengths (perimeter)	42 studs

16 + 16 + 5 + 5 or 2(16) + 2(5) = 32 + 10 = 42 studs (representing feet)

Other possible solutions include (but are not limited to):

11 x 10 studs: 11 + 11 + 10 + 10 = 2(11) + 2(10) = 22 + 20 = 42 studs

13 x 8 studs: 13 + 13 + 8 + 8 = 2(13) + 2(8) = 26 + 16 = 42 studs

18 x 3 studs: 18 + 18 + 3 + 3 or 2(18) + 2(3) = 36 + 6 = 42 studs

Remind students that the addition of all the side lengths creates a formula that works every time for every rectangular shape.

Problem #2: Build a rectangle with a perimeter of 50 inches in two different ways.

Have students build two different models with bricks (1 stud = 1 inch), then draw the models and prove that the perimeter of each rectangle is 50 inches. Have students record the side lengths of the rectangle in a table and write a math sentence for the solution.

Ask students: "What it is called when you use the same format to find the solution each time?" (*Answer*: formula)

Note: Students may put two baseplates together to build their models.

Possible Solution A:

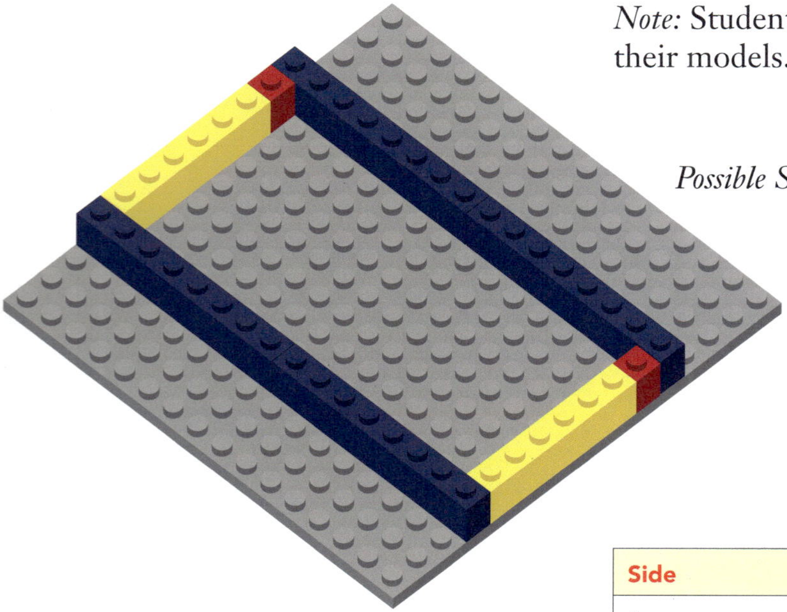

Side	Length in studs
1	16 studs
2	9 studs
3	16 studs
4	9 studs
Sum of all side lengths (perimeter)	50 studs

16 + 16 + 9 + 9 or 2(16) + 2(9) = 32 + 18 = 50 studs (representing inches)

Possible Solution B:

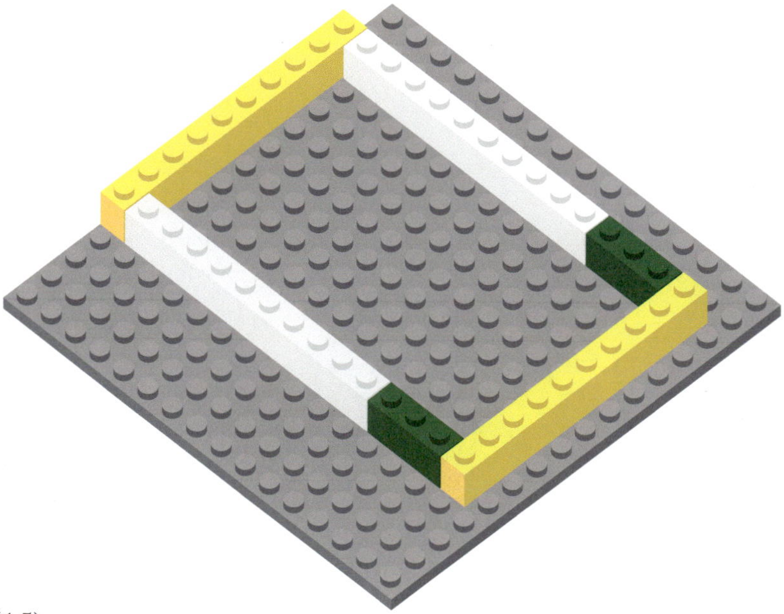

10 x 15 studs: 10 + 10 + 15 + 15 or 2(10) + 2(15) = 20 + 30 = 50 studs (representing inches)

Side	Length in studs
1	15 studs
2	10 studs
3	15 studs
4	10 studs
Sum of all side lengths (perimeter)	50 studs

Other possible solutions include (but are not limited to):
11 x 14 studs: 11 + 11 + 14 + 14 or 2(11) + 2(14) = 50 studs (or inches)

12 x 13 studs: 12 + 12 + 13 + 13 or 2(12) + 2(13) = 50 studs (or inches)

19 x 6 studs: 19 + 19 + 6 + 6 or 2(19) + 2(6) = 50 studs (or inches)

Problem #3: A rectangular swimming pool has a perimeter of 56 ft. How long is each side?

Have students build a models with bricks (1 stud = 1 foot), then draw the model and prove that the perimeter of the pool is 56 feet. Have students record the side lengths of the pool in a table and write a math sentence for the solution.

Ask students to identify the parts of the formula.

Possible solution:

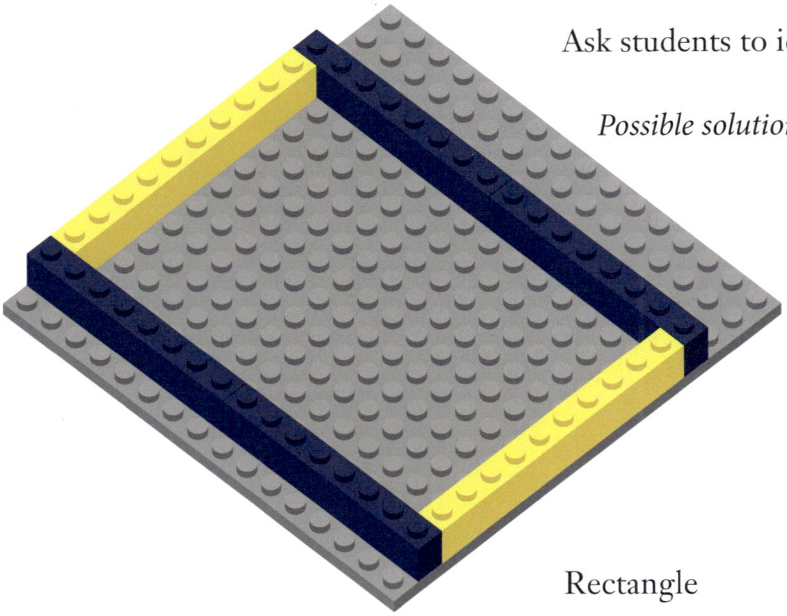

Rectangle

Side	Length in studs
1	16 studs
2	12 studs
3	16 studs
4	12 studs
Sum of all stud lengths (perimeter)	56 studs

16 + 16 + 12 + 12 or 2(16) + 2(12) = 32 + 24 = 56 studs (or feet)

Note: There are other possible solutions. One is a square swimming pool, measuring 14 x 14 studs:
14 + 14 + 14 + 14 or 4(14) = 56 studs (or feet)

Part 2: Show What You Know

1. The Jones family is building a rectangular patio. The total perimeter of the patio needs to be 36 feet. Design and build a model that meets the requirements for the patio. Draw your model and label the length of each side in studs. Write a math sentence to show that the total perimeter is 36 feet.

Two of several possible solutions:

9 + 9 + 9 + 9 or 4(9) = 36 studs (or feet)

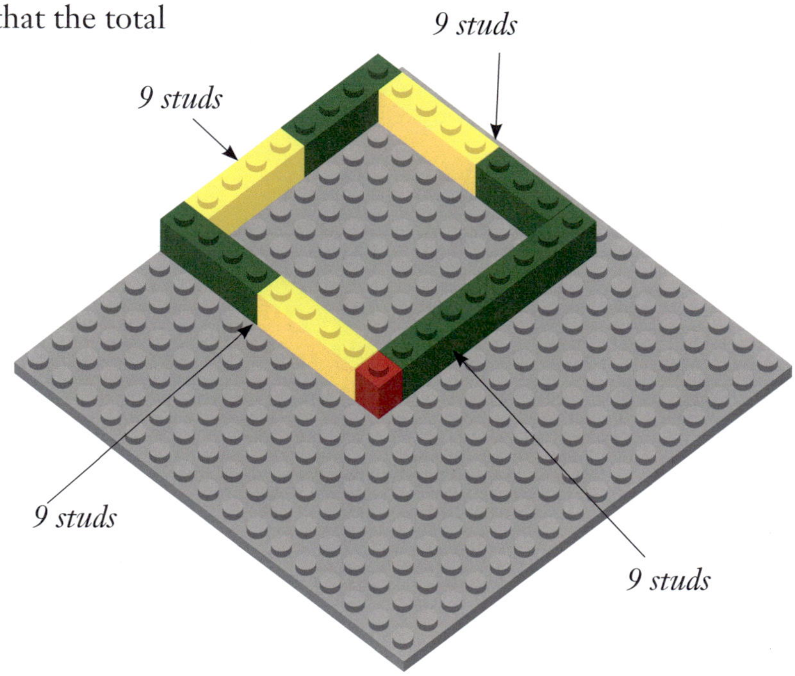

10 + 10 + 8 + 8 or 2(10) + 2(8) = 36 studs (or feet)

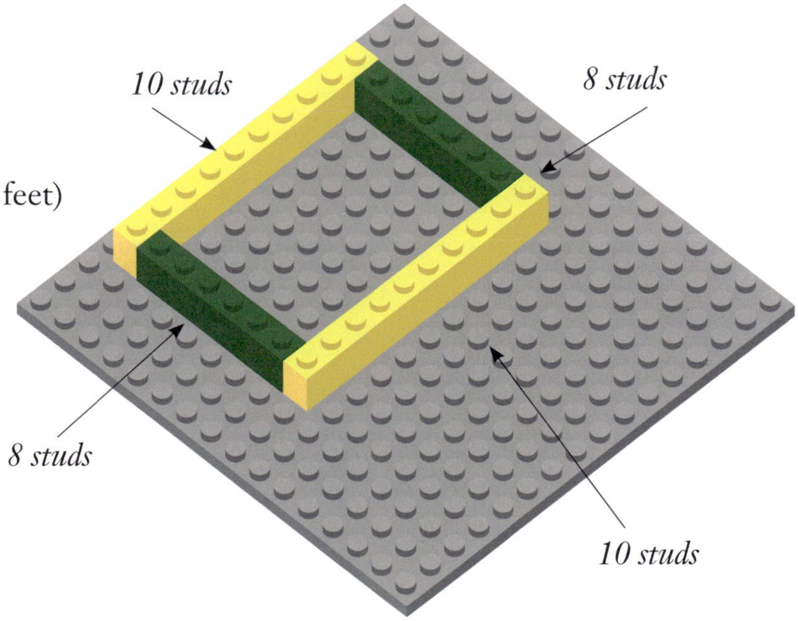

2. Amelia is putting a rectangular desk in her room. She has a space for a desk with a perimeter of 29 feet. If the desk has sides that are 6 feet, 6 feet, 8 feet, and 8 feet long, will it fit into the space? Build the outline of the desk to determine the perimeter, then compare it to the available space. Draw your solution and explain your thinking.

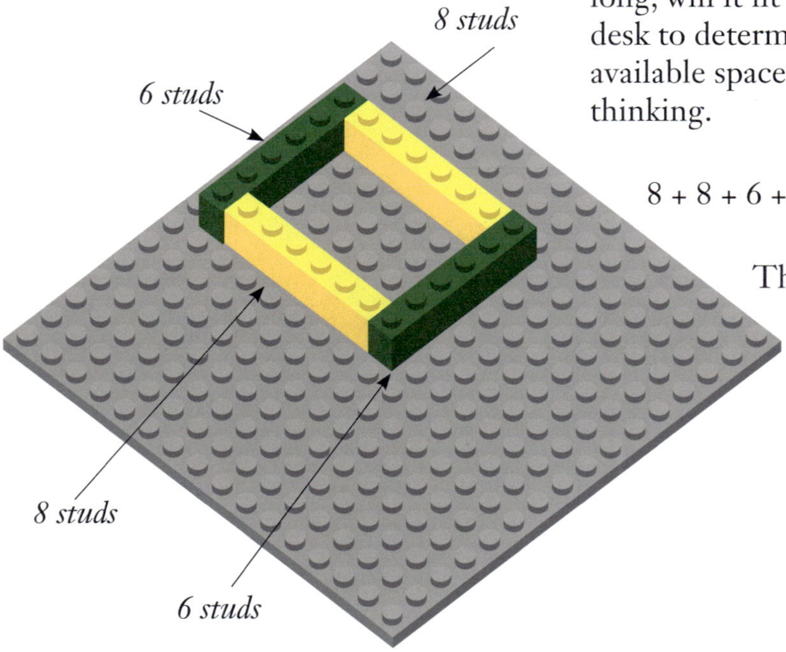

8 + 8 + 6 + 6 or 2(8) + 2(6) = 28 studs (or feet)

The space is big enough for the desk by one foot.

3. Jenna is creating a pen for a new rabbit. She has enough fencing material to make a perimeter of 34 feet. Design and build models of at least two solutions. Label the lengths of the sides of each model. Write a math sentence for each model that proves the perimeter is 34 feet. Identify the parts of the formula.

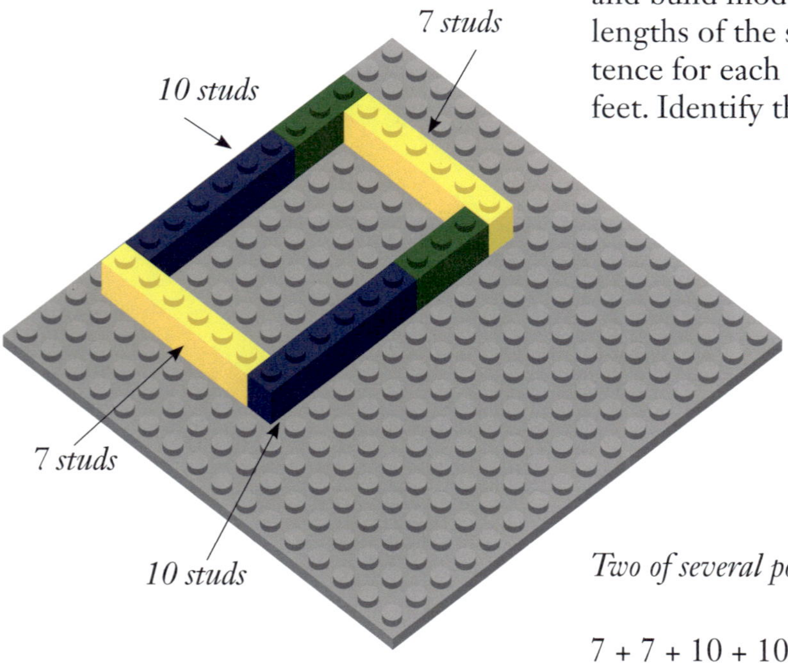

Two of several possible solutions:

7 + 7 + 10 + 10 or 2(7) + 2(10) = 34 studs (or feet)

$8 + 8 + 9 + 9$ or $2(8) + 2(9) = 34$ studs (or feet)

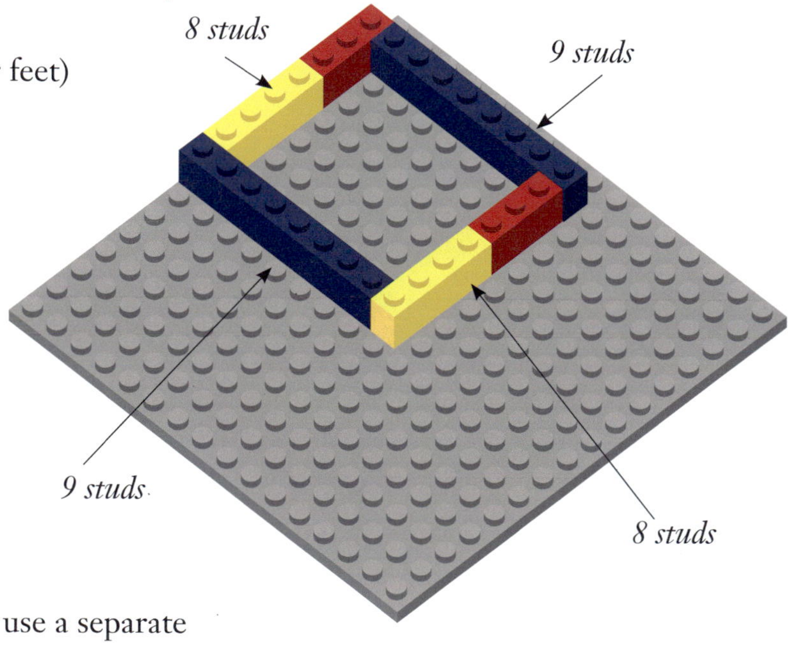

4. Partner Challenge: Each partner should use a separate baseplate to complete this challenge.

Build a rectangle that is 64 studs in perimeter. Inside the first rectangle, build another rectangle with a perimeter that is $\frac{1}{4}$ the size of the larger rectangle. Draw your model and explain your thinking. Write math sentences to show the perimeter of each rectangle.

Exchange models and prove that your partner's solution is correct. Discuss your findings with your partner. Identify the parts of the formula.

Solution:

Perimeter of large rectangle:
$16 + 16 + 16 + 16$ or $4(16) = 64$ studs

Perimeter of small rectangle:
$\frac{64}{4} = 4 + 4 + 4 + 4$ or $4(4) = 16$ studs

The large rectangle has a perimeter of 64 studs.

64 studs divided by 4 studs is 16 studs.

The small rectangle is 4 studs on each side.

8

SUGGESTED BRICKS

Size	Number
1x1	20
1x2	15
1x3	8
1x4	8
1x6	8
1x8	4
1x10	4
1x12	4
1x16	2
2x2	8
2x3	4
2x4	8
2x6	4
2x8	2
2x10	2

Note: Using a baseplate helps keep the bricks in place. One baseplate is suggested for these activities.

AREA

Students will learn/discover:

- How to determine the area of rectangular and square shapes
- How to use the mathematical formula for area

Why is this important?

A basic understanding of area is needed in everyday life. For instance, a decorator needs to calculate how much carpet or tile is needed to cover the floor of a room, or how much paint is needed to cover a wall. Calculating area is important in land use projects, such as gardening and fencing. Student should use squares and rectangles to learn that area is the amount of space or surface included within a set of lines. This topic should be explored as much as possible through real-world scenarios, not through the memorization of formulas. Before exploring this topic, students should have a firm understanding of perimeter (discussed in chapter 7).

Vocabulary:

- Area: The amount of space or surface included within a set of lines
- Exponent: A numeral that, when used in the context of measuring area, denotes the multiplication of the two dimensions of a shape (length x width); the exponent appears as a superscript 2 after the unit of measure (e.g., the area of the shape is 10 ft^2)
- Perimeter: The distance around the outside of a given shape; the sum of the measure of all the sides of a given shape
- Rectangle: A quadrilateral with 4 right angles and opposite sides equal
- Square: A quadrilateral with 4 right angles and all sides equal

How to use the companion student book, *Basic Measurement Using LEGO® Bricks–Student Edition*:

- After students build their models, have them draw the models and explain their thinking in the Student Edition. Recording the models on paper after building them with bricks helps reinforce the concepts being taught.
- Discuss the vocabulary for each lesson with students as they work through the Student Edition.
- Use the assessment in the Student Edition to gauge student understanding of the content.

Part 1: Show Them How

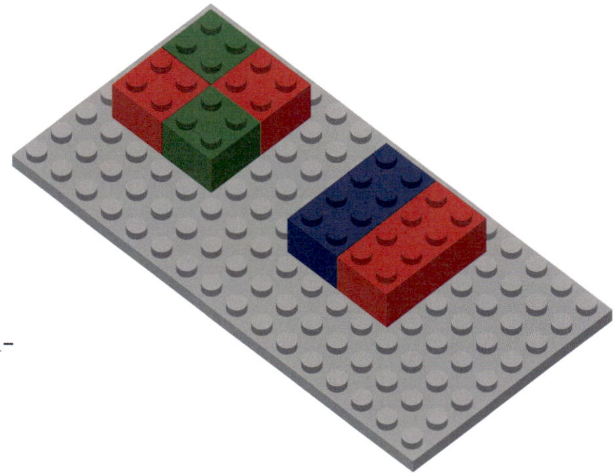

1. Have students build one array with four 2x2 bricks arranged in a square, and another array with two 2x4 bricks arranged in a square.

2. Review perimeter by asking students to count the number of studs on each side of one of the squares and add them together (*answer:* 16 studs).

3. Have students count the number of studs inside the square (*answer:* 16 studs). Explain to students that this is the *area* of the square in studs. *Note:* students should count all the studs of the square, including the studs they used to calculate perimeter, when calculating area.

Ask students what they notice (*answer:* the number of studs around the square is equal to the number of studs inside the square). Have students investigate whether this is true for every square, by creating squares of other sizes with bricks (*answer:* it is not true for every square).

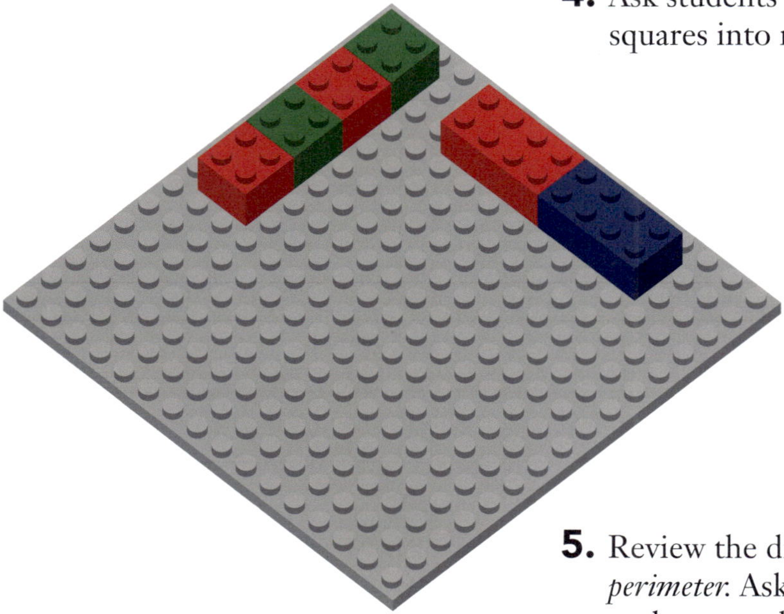

4. Ask students to arrange the bricks from the two original squares into rectangles that are not squares.

5. Review the difference between the terms *area* and *perimeter*. Ask students to find the perimeter and area of each rectangle in studs (*answer:* the perimeter is 20 studs and the area is 16 studs). Explain that rectangles do not always have the same area (number of studs inside) as they do perimeter (number of studs around the outside).

6. Ask students to find another way, besides counting, to determine how many studs are on the inside of each rectangle (*answer:* multiply the number of studs on the shorter side by the number of studs on the longer side).

Explain that the longer side, denoted by "*L*," is the length of the rectangle. The shorter side, denoted by "*W*," is the width of the rectangle. Point out that area is referred to as "squared" because it is the product of two dimensions ($L \times W = A^2$). Show students how to write area as a square unit of measure using an exponent (*examples:* ft^2, in^2, cm^2, km^2) and how to refer to the square units (*examples:* square feet, square inches, square centimeters, square kilometers).

Have students practice building rectangles with different lengths and widths, using multiplication to find the area. For example, ask students to build a rectangle of 3 studs x 4 studs and find its area (*answer:* three 1x4 bricks; area = 3 studs x 4 studs = 12 $studs^2$). *Note:* remind students that a 3x4 rectangle is equivalent to 3 sets of 4 studs (see *Multiplication Using LEGO®Bricks—Teacher Edition*).

Possible solution:

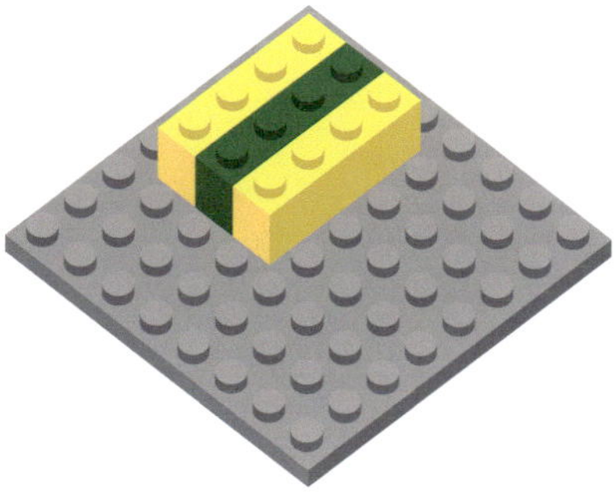

L = 4   W = 3
Area = 12 square studs (4 studs x 3 studs = 12 studs^2)

Problem #1: Josh is building an outdoor pen for chickens. The pen must fit in his backyard and it needs to be rectangular. The area of the pen can only be 32 ft^2.

Have students model the pen with bricks (1 stud = 1 foot), then draw the model and prove that the area is 32 ft^2.

Have students write a math sentence for the solution.

Possible solution:

L = 8   W = 4
Length = 8 studs (or feet)
Width = 4 studs (or feet)
Area = 8 x 4 = 32 studs^2 (or ft^2)

Problem #2: Model a rectangle with an area of 24 in^2 in two different ways (1 stud = 1 inch).

Have students draw the models and prove each solution by writing a math sentence for each solution.

Note: There are many solutions to this problem. Have students experiment and share their solutions. Students can also make a table to record their solutions.

Possible solutions:

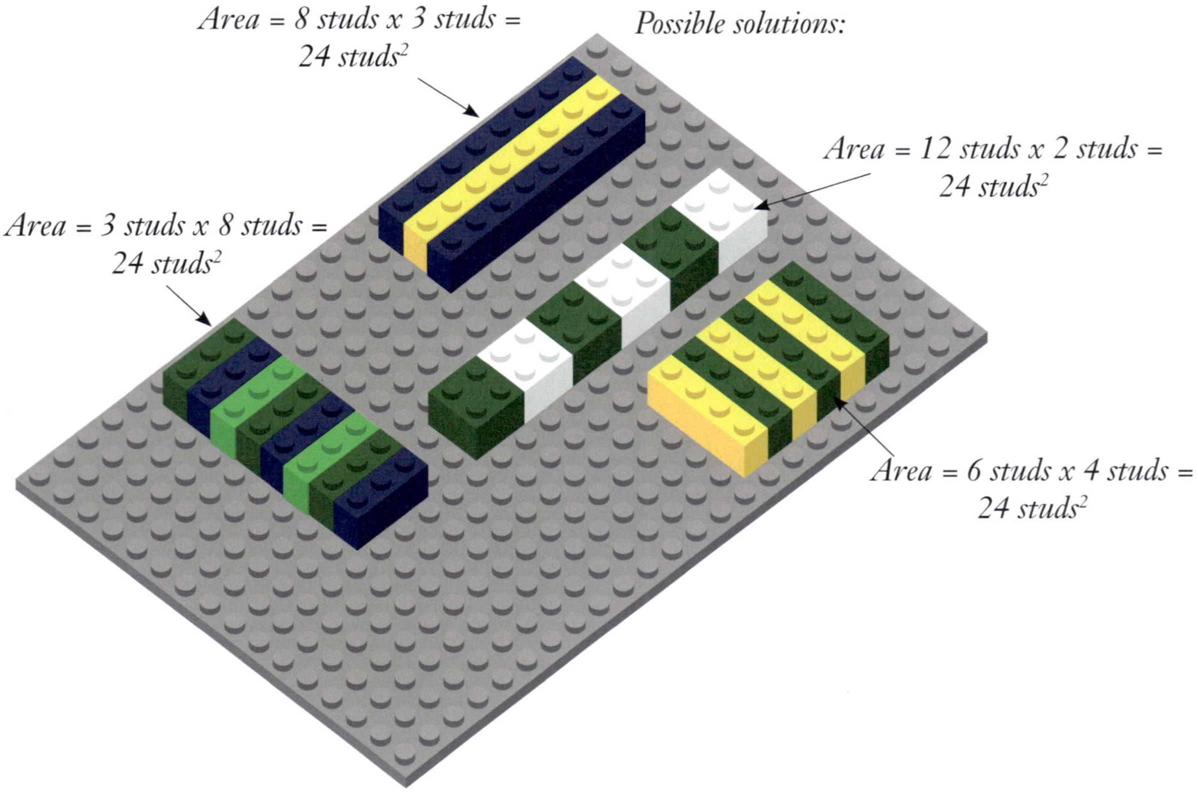

Problem #3: A square swimming pool has an area of 36 ft^2. How long is each side of the pool?

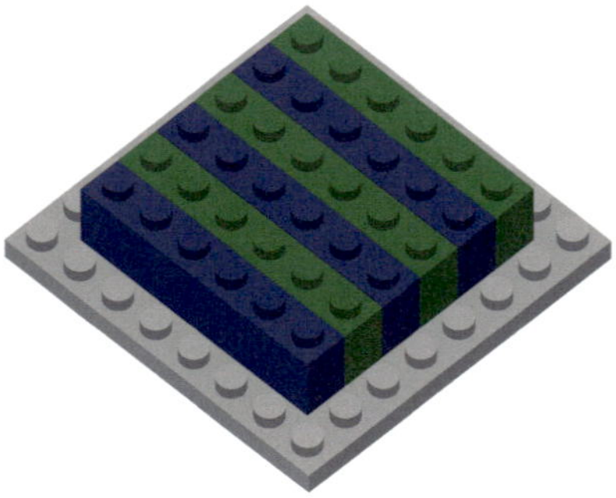

Have students model the pool with bricks (1 stud = 1 foot), then draw the model and prove that the area is 36 ft^2.

Have students write a math sentence for the solution.

Solution:

Area = 6 x 6 ft = 36 ft^2
Length per side = 6 ft

Part 2: Show What You Know

1. The Jones family is building a rectangular patio. The total area of the patio needs to be 56 ft^2. Can you design and build a model that meets the requirements for the patio? Draw your model and label the length and width in studs. Write a math sentence to show that the total area is 56 ft^2.

Possible solution:

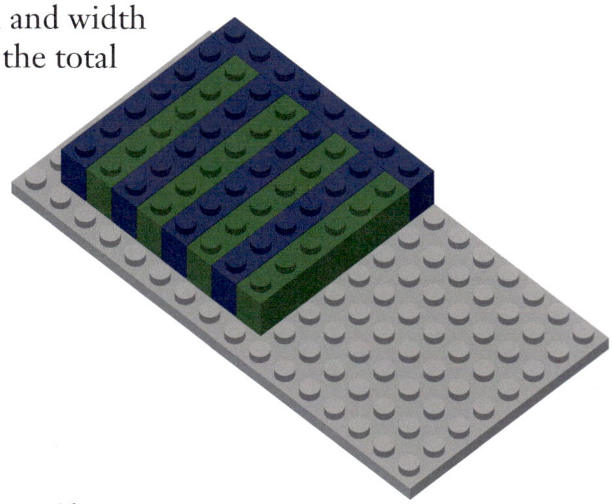

Length = 8 studs
Width = 7 studs
Area = 8 x 7 = 56 ft^2

2. Amelia is putting a rectangular rug in her room. She has a space for a rug with an area of 28 ft^2. If the rug has sides that are 4 feet and 8 feet long, will it fit into the space? Build a model of the rug to determine the area, then compare it to the available space. Draw your solution and explain your thinking.

Possible solution:

The rug will not fit, because the area of the rug is 4 ft^2 larger than the area of the available space.

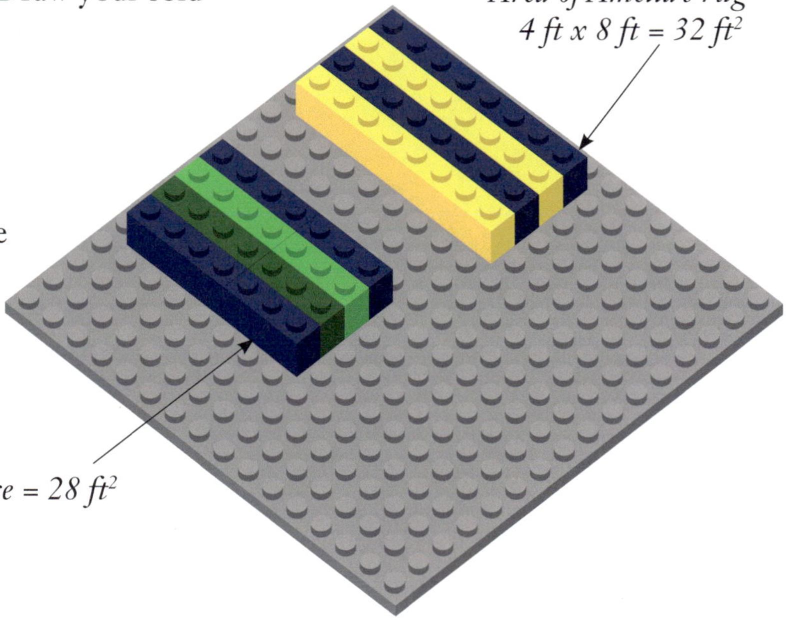

3. Logan is creating a stall for a new horse. She wants the stall to have an area of 306 ft^2. Build a model for the stall. Label the length and width of your model. Write a math sentence that proves the area is 306 ft^2.

Note: Two smaller baseplates or one large baseplate will be needed to build this model.

Possible solution:

Length = 18 studs
Width = 17 studs
Area = 18 x 17 = 306 ft^2

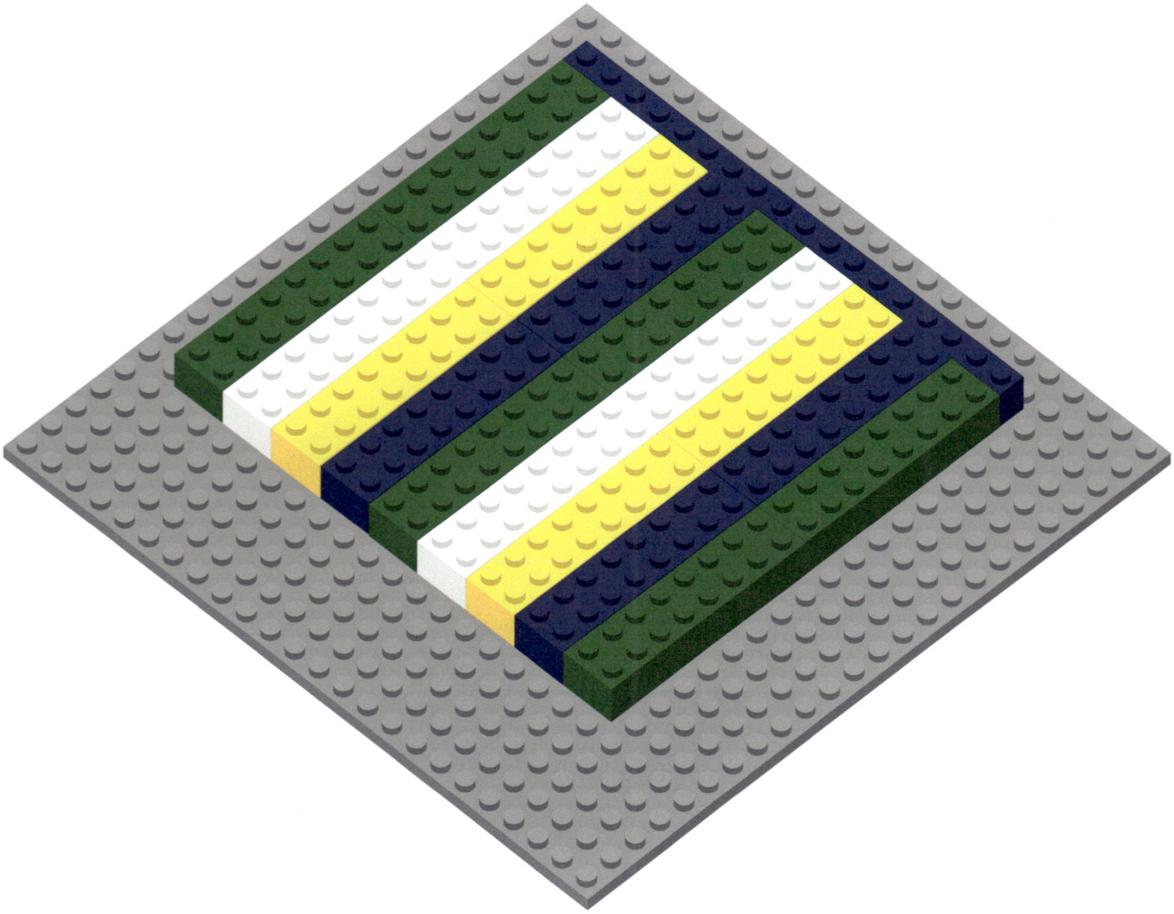

4. Write your own problem and model the solution. Share your problem with a partner, but do not show your partner the solution. Solve each other's problems, then compare solutions.

Solutions will vary.

APPENDIX

- **Suggested Brick Inventory**
- **Student Assessment Chart**
- **Baseplate Paper**

SUGGESTED BRICK INVENTORY

SIZE	NUMBER
1x1	84 (32 each of two colors and 10 each of two more colors)
1x2	25 (10 each of two colors and 5 of a third color)
1x3	12 (6 each of two colors)
1x4	10
1x6	10
1x8	6
1x10	6
1x12	5
1x16	2
2x2	12
2x3	6
2x4	9
2x6	4
2x8	2
2x10	2

BASIC MEASUREMENT

Student Assessment Chart

Name ____

Performance Skill	Not yet	With help	On target	Comments
I can identify standard units of measure (customary/metric).				
I can compare measures using different units.				
I can model coin values up to one dollar, using bricks.				
I can add coin values (pennies, nickels, dimes, and quarters) within one dollar.				
I can tell time to the nearest hour, half hour, and 5-minute intervals.				
I can determine elapsed time.				
I can determine the number of quarts, pints, and cups in a gallon.				
I can identify the units in the metric system.				
I can describe how the metric system is related to the base ten system.				
I can relate place value to reading decimal numbers.				
I can make basic conversions using the metric system.				
I can use a model to find the perimeter of rectangular shapes.				
I can use the mathematical formula for perimeter.				
I can use a model to determine the area of rectangular shapes.				
I can use the mathematical formula for area.				

BASEPLATE PAPER

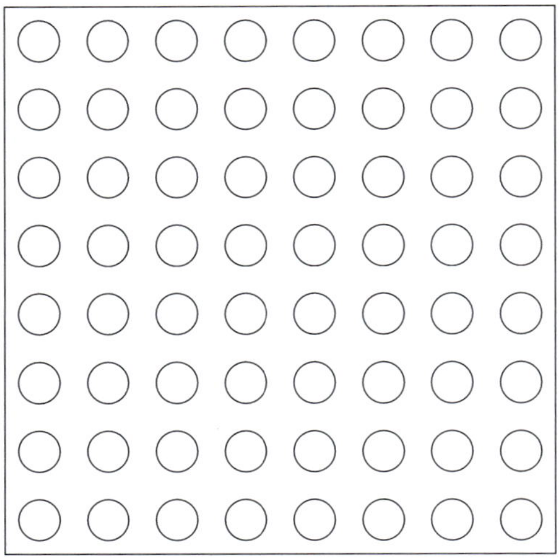

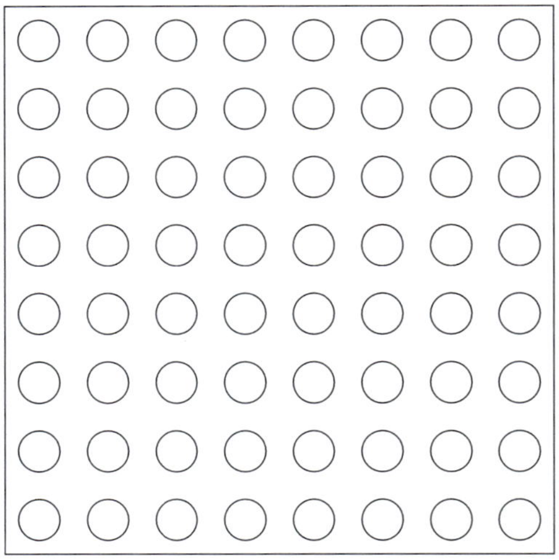

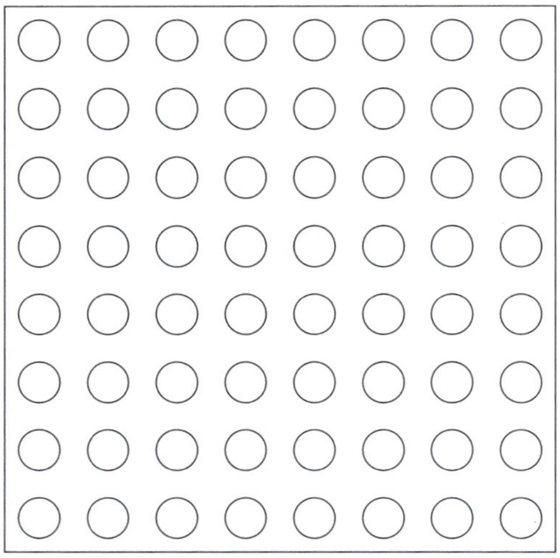

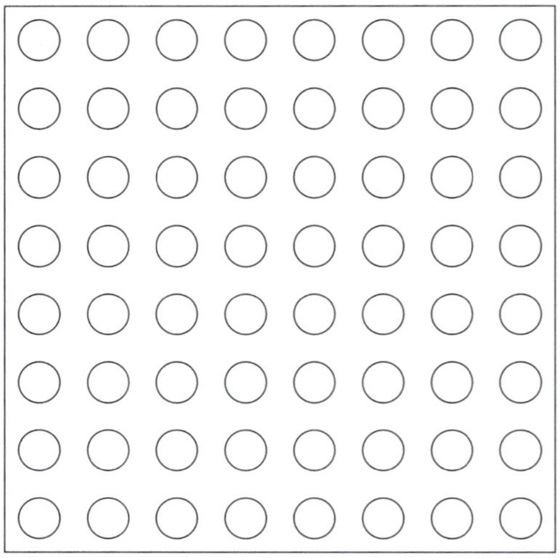

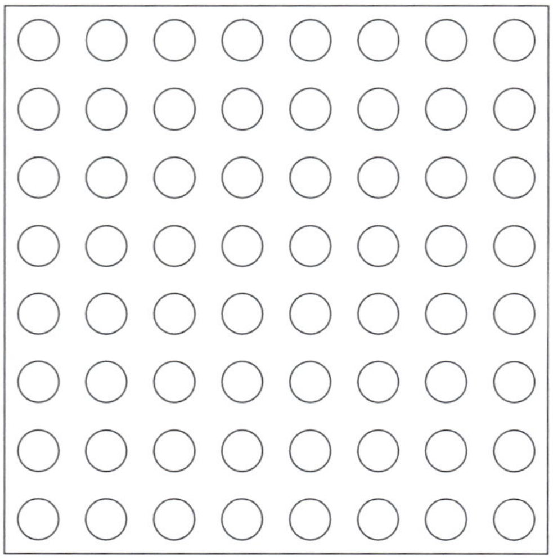

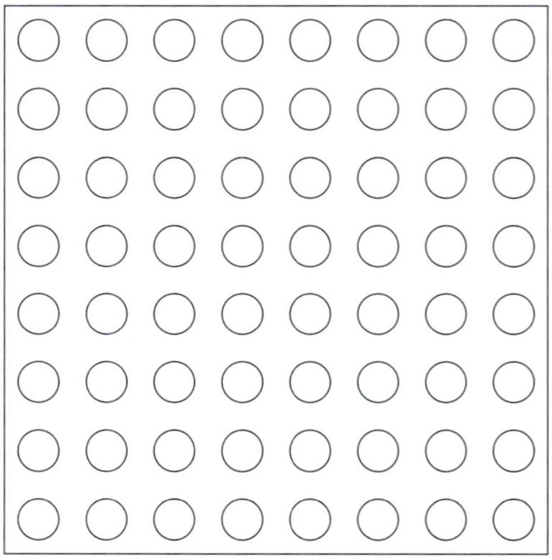

BASEPLATE PAPER

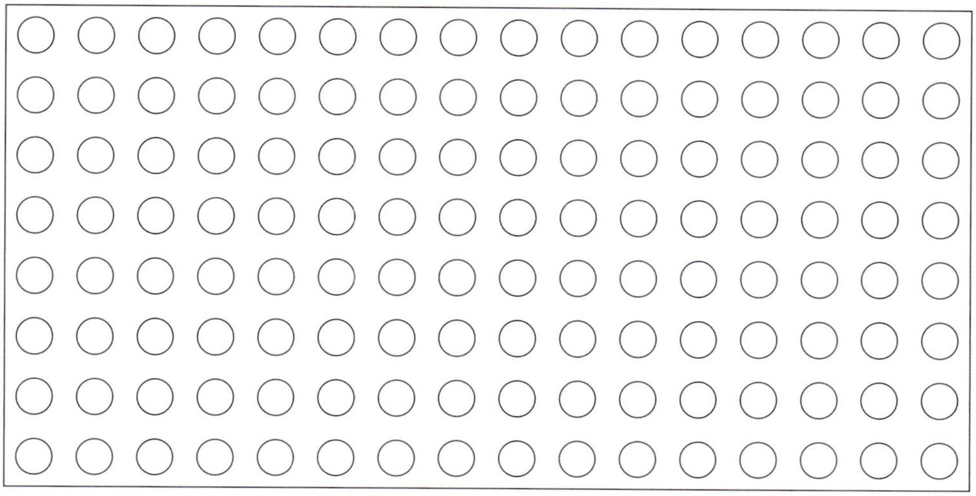

Made in the USA
Las Vegas, NV
03 November 2026

57422365R00052